Lecture Notes in Control and Information Sciences 445

For further volumes:
http://www.springer.com/series/642

Dong Wang, Peng Shi, and Wei Wang

Robust Filtering and Fault Detection of Switched Delay Systems

Springer

Dr. Dong Wang
School of Control Science and Engineering
Dalian University of Technology
Dalian
Peoples Republic of China

Prof. Wei Wang
School of Control Science and Engineering
Dalian University of Technology
Dalian
Peoples Republic of China

Prof. Peng Shi
College of Engineering and Science
Victoria University
Melbourne
Australia

and

School of Electrical and Electronic Engineering
The University of Adelaide
Adelaide
Australia

ISSN 0170-8643 ISSN 1610-7411 (electronic)
ISBN 978-3-642-37684-9 ISBN 978-3-642-37685-6 (eBook)
DOI 10.1007/978-3-642-37685-6
Springer Heidelberg New York Dordrecht London

Library of Congress Control Number: 2013935587

Printed on acid-free paper

Springer is part of Springer Science+Business Media (www.springer.com)

Preface

Switched systems, a class of multi-model systems, are composed of both continuous dynamics and discrete dynamics. The continuous dynamics is characterized by the continuous changes of the system dynamic such as continuous-time or discrete-time subsystems. The discrete dynamics is a switching law specifying which subsystem is activated. Such systems arise widely in real engineering applications. First, a couple of dynamical subsystems are required to model an engineering system due to the abrupt parameter variations, networks with periodically varying switchings and sudden change of system structures. Second, the multi-controller switching scheme can present the good control performances of the systems in the presence of large uncertainties or disturbances when no single controller is effective. Therefore, switched systems have been attracting great attention in the control and system community during the past two decades. A great number of books and papers have been published with the development of the relevant concepts and powerful tool. In the existing results, the issues are studied in the following three types of switching signals: (1) arbitrary switching signals; (2) certain kinds of switching signals; (3) designed switching signals. The former two classes are used to analyze problems when the switching signals are available while the third is to construct the switching signals to provide the solutions of the issues such as stabilization.

On the other hand, time delay is universal in real applications because of the signal transmission in communication, electronics, hydraulic and chemical systems, mechanical structures and so on. The existence of time delays in a system can usually degrade the system performance and even cause the instability. A number of research efforts have been done to develop some approaches such as descriptor systems approach, Newton-Leibniz formula, bounding inequalities, free-weighting matrix approach and so on. The results reported can be classified into two types: delay-independent results applying to arbitrary size of time delays and delay-dependent results containing the information on the size of delays. Generally speaking, the delay-dependent results are less conservative than the delay-independent ones when the size of the time delay is small. Considering the characteristic of switched systems, there exist two ways involving time delays: switching signals and subsystems, which are called switched delay systems. The application of switched delay systems can be widely found in engineering such as Networked Control Systems and power systems. Due to the interaction among the continuous-time (discrete-time) dynamics, switching signals and the existence of time delays, the dynamic behavior of switched delay systems is very complicated. Mechanism of system evolution is not clear and many problems of analysis and synthesis of such systems require to be studied. For example, in real industrial process systems, the disturbances and parameter uncertainties are inevitable. Based on the measurement output of the system, different filters, depending on the type of noises, are designed to estimate the almost true signals of the systems, thus robust filtering has become a hot topic in recent few years. Meanwhile, the systems operating in the normal cases or not directly affect the economic interests and are related to people's lives. If the faults of the system are not able to be promptly detected and removed, it may cause the failure and paralysis of the systems, the huge losses of the production, and even lead to catastrophic consequences. In order to meet the requirements for the high safety and reliability of industrial process systems, fault diagnosis and fault tolerance of dynamic systems have received great attention. Fault detection, the first and crucial step of fault diagnosis, is employed to monitor the occurrence of a fault. Although many results on the filter design and fault detection have been reported so far, there lacks a systemic way to deal with such problems, which motivates the writing of the current monograph.

This book studies the problems of filter design and fault detection of switched delay systems including two parts. In the first part, the problem of filter design for switched systems with state delays is investigated under different switching signals. First, a novel linearization technique is proposed, combining with the switched Lyapunov function method, delay-independent sufficient conditions for the existence of filters are obtained. Then, by using free-weighting matrix method, delay-dependent sufficient conditions for the existence of filters are established. The results mentioned above are concerned with the arbitrary switching signals. Furthermore, by introducing a proper factor to construct a novel Lyapunov function and using average dwell time (ADT) approach, sufficient conditions for the solvability of this problem are obtained. In the second part, the problems of robust fault detection for switched systems with state delays under an arbitrary switching signal and a switching signal with ADT are studied. The fault detection filter is employed as the residual generator and designed to minimize the estimation error between the residuals and faults. By utilizing H_∞ theory, sufficient conditions for the solvability of this problem are provided. Finally, some perspectives and future extensions of switched delay systems are presented.

This book is helpful to learn and understand the fundamental concepts and schemes of switched delay systems and fault detection. The main objective of this book is to provide a primary reference book for researchers and engineers in the system and control community. It can also serve as complementary reading for the postgraduate in the field of switched systems.

We would like to express our gratitude to our parents and families who have given the great support to write this book. Dong Wang is indebted to his parents, Fenglin Wang and Shuqing Zhang, and his wife, Jie Lian for their understanding and encouragement throughout these years. Wei Wang would like to thank his wife, Zili Zhan, for encouragements and enthusiastic support. Finally, we would like to thank the team of Springer for their cooperation and encouragement in bringing out the work in the form of monograph.

Dalian University of Technology, China — Dong Wang
Victoria University; and The University of Adelaide, Australia — Peng Shi
Dalian University of Technology, China — Wei Wang

Feb. 2013

Acknowledgements

The contents contained in this book are an outgrowth of our academic research activities over the past several years. The authors are grateful to Dalian University of Technology for providing a great number of resources for our research work. We would like to acknowledge the support of research grants, including National Natural Science Foundation of China (Grants No. 61104114, 61004040, 61004020), China Postdoctoral Science Foundation (Grant No. 2012T50254), Fundamental Research Funds for the Central Universities, the Engineering and Physical Sciences Research Council of UK (EP/F029195). The Published book is sponsored by the Dalian Municipal Government.

Contents

1 Introdution

This chapter briefly describes some basic concepts and tools for studying the relevant topics of switched delay systems. In addition, the preview of each chapter is provided.

1.1 Switched Delay Systems

During the past two decades, switched systems have been widely studied and many interesting results have been reported in the literature, for instance, [65, 67, 11, 111, 112, 99, 6, 42, 10, 88] and the references therein. Such systems consist of a family of continuous-time or discrete-time subsystems and a switching signal specifying which subsystem is activated. This means that a switched system admits continuous states taking values from a vector space and discrete states from a discrete index set. The interaction between continuous and discrete states renders such systems complex. In spite of the linear subsystems, switched linear systems can represent very complicated behaviors bridging the linear systems and the complex dynamical systems. In the existing results [65, 67], the issues are studied in the following three types of switching signals:

1. Arbitrary switching signals;
2. Certain kinds of switching signals;
3. Designed switching signals.

The former two classes are used to analyze stability problems when the switching signals are available while the third is to construct the switching signals to provide the solutions of the issues such as stabilization. The motivation to study switched systems is mainly in twofold. First of all, many engineering systems can be represented by switched systems, such as Networked Control Systems (NCS) [59] and [69], automotive engine control and aircraft control [2]. Secondly, the idea of controller switching is introduced in order to overcome the shortcomings of the single controller and improve system performance [78, 43, 50]. As is well known that the existence of Lyapunov function candidates is key in the analysis of systems. The

D. Wang et al.: *Robust Filtering & Fault Detect. of Switched Delay Systems*, LNCIS 445, pp. 1–19.
DOI: 10.1007/978-3-642-37685-6_1

internal stability of the system under consideration is concluded if a relevant Lyapunov function is found. For switched systems, it is a simple and direct method to find the common Lyapunov function for all the subsystems [83, 79, 97, 66, 98]. A great number of techniques have been proposed to construct such a Lyapunov function [11, 67, 83]. However, it turns out that most switched systems in practice do not possess a common Lyapunov function, especially for nonlinear switched systems. Though there is no common Lyapunov function, switched systems may still be asymptotically stable under some properly switching law. At this point, one has to consider the switching rules from two respects. One is that the switching laws are unknown but fixed and available online for designers. In this case, stability conditions of the switching system under these switching laws are the objective of the research. The other is that the switching law is considered as a design variable and then constructed to stabilize the systems. So far, these two cases are studied developing single Lyapunov function approach [67], multiple Lyapunov functions approach [6], average dwell time technology [42], switched Lyapunov function approach (SLF) [10] and so on. We will provide more details of the following approaches relevant to this book.

1.1.1 Single Lyapunov Function Approach

The main idea of single Lyapunov function approach is to construct such a Lyapunov function for each system that each system is stable in the corresponding region [11, 118, 117, 101]. For example, suppose there are N subsystems in a switched system. The first step is to partition the entire state space R^n into N subspaces corresponding each subsystem and no common region among these subspaces. That is, $R^n = \Omega_1 \cup \Omega_2 \cup \cdots \Omega_N$, $\Omega_i \cup \Omega_j = \emptyset$, $i \neq j$. The second step is to choose Lyapunov function. When the system state enters into the region Ω_i, the *ith* subsystem is active and the corresponding Lyaupnov function decreases. When the above two steps are accomplished, switched systems are stable in the entire space. It is seen from these two design steps that the switching law is state-dependent. The switching controllers are designed using the single-Lypaunov function method together with convex combination technique. The disadvantage of such method is the difficulty of selecting a Lyapunov function which decreases in every partitioned region

1.1.2 Multiple Lyapunov Functions Approach

The multiple Lyapunov function technique proposed by Peleties and DeCarlo [88] and later generalized by Branicky [6] and Hou, Michel and Ye [45, 135] is a product of the traditional Lyapunov function method applied to switched systems. The key point of the method is the non-increasing requirement on any Lyapunov function over the 'switched out' ('switched on') time sequence of the corresponding subsystem. Specifically, the Lyapunov-like function is selected for each subsystem. The value of the Lyapunov-like function at the end (the beginning) of the next running intervals is less than that at the end (the beginning) of the current running intervals, then, the energy of the whole system shows a decreasing trend. This means that the systems are asymptotically stable. Zhao introduced weak Lyapunov-like functions and then derived a necessary and sufficient condition for stability in terms of multiple Lyapunov functions, which generalizes Branicky's condition and Hou, Michel and Ye's condition [148]. The principle of multiple Lyapunov function is shown in Fig. 1.1. The symbols $\triangle, \circ$ and $\square$ represent the values of Lyapunov functions V_1, V_2 and V_3 for three subsystems on the 'switched on' time sequence $0, t_1, \ldots, t_6$. The growth of Lyapunov-like functions associated with three subsystems is bounded on the running time intervals of the corresponding subsystem such as $[0, t_1)$ and $[t_5, t_6)$. It is observed that all the symbols $\triangle, \circ$ and $\square$ satisfy the non-increasing conditions, which implies switched systems under such switching law are asymptotically stable.

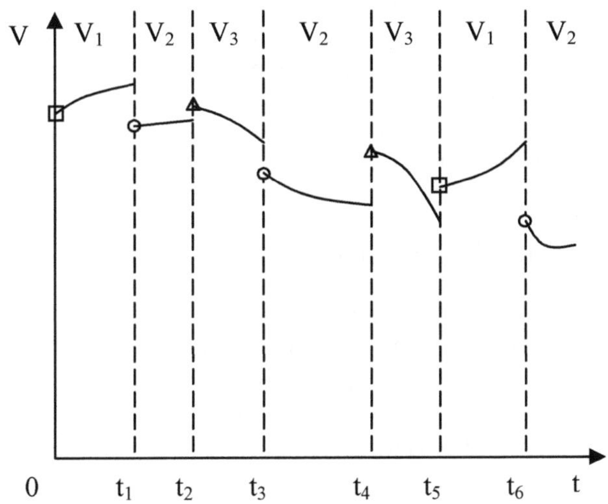

Fig. 1.1 The principle of multiple Lyapunov function

1.1.3 Average Dwell Time Technology

The proposed weak Lyapunov-like function associated with a subsystem relax the requirement that each Lyapunov function must decrease on the time intervals when the subsystem is active, namely, the bounded growth is allowed. It requires the values of Lyapunov function at the switching time instants, which involves the knowledge of the states. This is usually hard to satisfy and check. As a matter of fact, in order to apply multiple Lyapunov function method, connecting adjacent Lyapunov functions at switching points is an accepted strategy. This is achieved by choosing the switching law according to the 'min-switching' strategy of all Lyapunov functions [67], which is a special case of Branicky's non-increasing condition, but easy to design and realize. It is well known that a switched system is stable if each subsystem is stable and their active interval is enough long to dissipate the transient effect after each switching. Morse introduced the notions of dwell time to describe sufficiently long active interval [78]. This is also a slow switching idea that is to introduce a constant $\tau > 0$ and restrict a switching signal to satisfy $t_{i+1} - t_i > \tau$, where t_1, t_2, $\cdots$ are the switching time instants and i is the integer. The proof of stability is given by using multiple Lyapunov function. In the stabilization context, specifying a dwell time is conservative. For instance, when the controller for stabilizing the systems fails temporarily, if a dwell time is still carried out, then, the performance may deteriorate to an unacceptable level. It is possible to activate the controller for enough long time in the subsequent switching sequence to compensate the performance damage and achieve the acceptable level. Based on this idea, the notion of average dwell time switching is proposed [42], which specifies that the time interval of switching signals between consecutive switchings is at least τ^* on the average. The average dwell time method is extended to study the stability of switched systems with stable and unstable subsystems simultaneously [138, 137]. It is shown as long as these unstable subsystems are activated in a relatively short time, the stability of switched systems can be still guaranteed. The idea of average dwell time turns out to be very useful in stability analysis of switched and is extensively applied [139, 114, 104, 142, 105].

1.1.4 Switched Lyapunov Function Approach

Another important kind of Multiple Lyapunov function is switched Lyapunov function proposed by Daafouz [10]. Such method sufficiently utilizes the discrete time property. Besides in the running time interval, the values of Lyapunov function still decrease at the switching points. The principle is indicated in Fig. 1.2. Switched Lyapunov function has been employed widely for stability analysis and synthesis of switched systems [70, 71, 15, 71, 127]. It is important to note that common Lyapunov function will no longer be necessary and sufficient conditions to ensure the stability of switched systems under arbitrary switching signals because arbitrary switching signals are also given using switched Lyapunov function.

On the other hand, a time delay results from long transmission lines, finite speed of information processing rate and causal nature of systems. It is also universal in real applications i.e., the signal transmission in communication, electronics, hydraulic and chemical systems, mechanical structures and so on [93, 17, 37, 56]. Time delay systems means that the future evolution of the state variables in the systems depend on both the current and the past values. To describe such kind of systems, delay differential equations (DDE) are employed instead of the conventional ordinary differential equations (ODE). It is well-known that time delays are the inherent features of many physical process and the main sources of poor performances and instability [37, 27, 3]. Two basic stability theorems are used to present the stability conditions: Lyapunov-Razumikhin Theorem and Lyapunov-Krasovskii

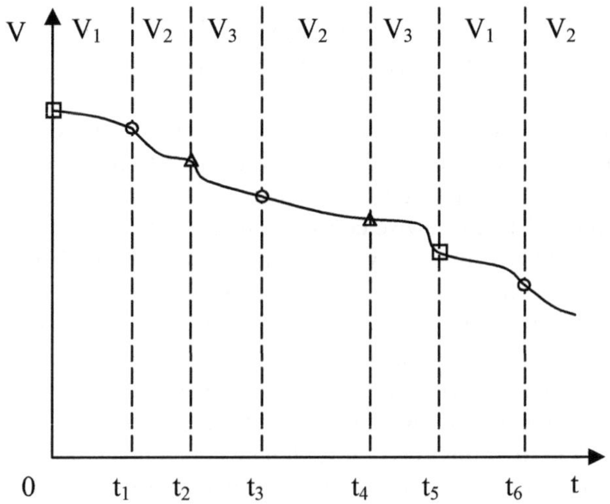

Fig. 1.2 The principle of switched Lyapunov function

Theorem [57]. The former constructs Lyapunov function while the latter the Lyapunov functionals. Although the latter renders the analysis and synthesis more complicated than the former, it presents less conservative results, which is the objective of most of the researchers in the recent years. Furthermore, some approaches are developed to reducing the conservatism combining with Lyapunov-Krasovskii Theorem such as descriptor systems approach, Newton-Leibniz formula, bounding inequalities, free-weighting matrix approach and so on. Based on different approaches, the results obtained can be classified into two types: delay-independent and delay-dependent [29, 33, 24, 76, 123, 25, 85]. Since the delay-independent conditions do not include the information of time delays, it is applicable to arbitrary size of time delays. The delay-dependent conditions contain the information on the size of delays. Generally speaking, the delay-dependent results are less conservative than the delay-independent ones when the size of the time delay is small.

Considering the characteristic of switched systems, there exist two ways involving time delays: switching signals and subsystems. More specifically, time delays can be introduced by state delays and/or input/output delays. The application of switched systems with state delays can be widely found in engineering such as Networked Control Systems [136, 115, 126, 69]. Hence, switched systems with state delay have gained a great deal of attention (see, for example, [111, 112, 44, 15, 58, 144] and the references therein), which is also the main research plant of the book. Taking as an example switched linear systems with state delays, such systems are mathematically described by

$$\delta x(t) = A_\sigma x(t) + E_\sigma x(t - d(t)) + B_\sigma u(t),$$
$$y(t) = C_\sigma x(t) + F_\sigma x(t - d(t)) + D_\sigma u(t), \quad (1.1)$$
$$x(\theta) = \phi(\theta), \quad \theta \in [-h_2, 0],$$

where δ denotes the derivative operator in the continuous-time context (i.e., $\delta x(t) = \dot{x}(t)$) and the shift forward operator in the discrete-time context (i.e., $\delta x(t) = x(t+1)$), $x(t) \in R^n$ is the state, $y(t) \in R^r$ is the measured output and $u(t) \in R^s$ is the control input. The piecewise constant function σ denoting $\sigma(t)$ for simplicity : $[0, \infty) \to \mathscr{P} = \{1, \cdots, p\}$ is a switching signal (a switching law) to specify, at each time instant t, the index $\sigma \in \mathscr{P}$ of the active subsystem and $p > 1$ is the number of subsystems. $d(t)$ denotes a time delay, which is allowed to be a constant delay ($d(t) = h_2$), a time-varying delay ($0 < d(t) \leq h_2$), an interval time-varying delay ($h_1 \leq d(t) \leq h_2$) and so on, where $h_2 > h_1 \geq 0$ are constant. The third equation in (1.1) is the initial condition which is given for the well-defined solution of the first equation in (1.1). It is obvious that such a switched delay system is a multi-model system, i.e., $\sigma = i$ means that the ith subsystem represented by constant matrices A_i, E_i, B_i, C_i, F_i and D_i is activated. A diagram of switched delay systems is given in Fig. 1.3.

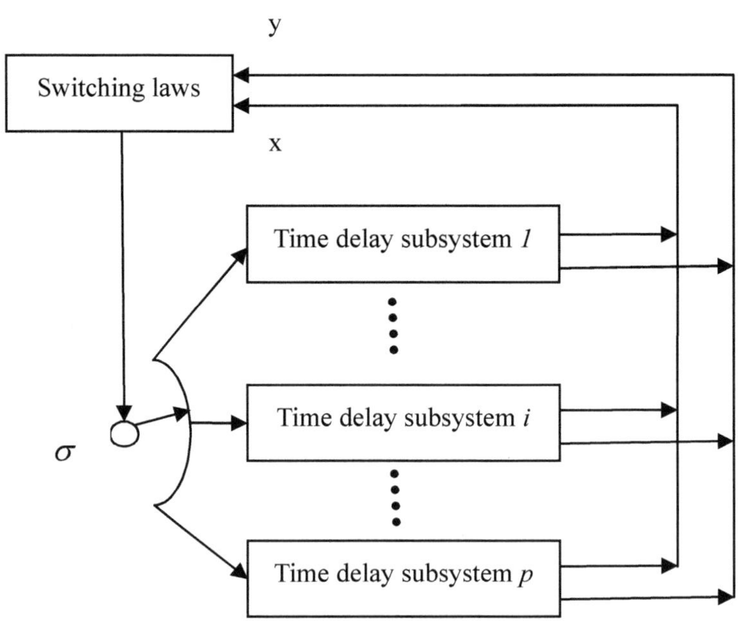

Fig. 1.3 A diagram of switched delay systems

1.1.5 A Motivating Example

The problem of preserving the water quality constituents to standard levels in multi-reach fresh water streams is investigated in [72], where a multi-model representative of water pollution in multi-reach fresh water streams of the River Nile are presented. The model of such system is described by

$$\begin{aligned} x_{k+1} &= A_{\sigma}x_k + A_{d\sigma}x_{k-\tau_0} + B_{\sigma}u_k + D_{1\sigma}w_k, \\ z_k &= G_{\sigma}x_k + G_{d\sigma}x_{k-\tau_0} + D_{2\sigma}w_k, \\ u_k &= \beta_k \hat{u}_{k-\tau_1}, \\ y_k &= \alpha_k C_{d\sigma}x_{k-\tau_2} + D_{3\sigma}w_k, \\ x_k &= \phi_k,\ k = -\tilde{d},\ -\tilde{d}+1,\ \cdots,\ 0,\ \tilde{d} = max\{\tau_0,\ \tau_1,\ \tau_2\}. \end{aligned} \tag{1.2}$$

where the state x_k denotes water quality constituents such as algae and ammonia nitrogen. The disturbance w_k arises from irregular discharge of effluents. Time delay τ_0 represents an average time to clear water streams up, which reflects the mixing effect of biochemical constituents in the reach. By means of the controller design, the water quality constituents are remained in the standard level. Three subsystems $i = \{1,\ 2,\ 3\}$ are set up to represent three operating points, each of which depicts the influence of environmental factors. The switching signal σ is not known a prior but its instantaneous value is available online by water pollution management. The switching among these subsystems is dependant of the industrial dumped discharges followed by irregular patterns. The control signals are generated by modifying water speed or discharging an amount of effluents.

As the measurement devices in the river are far from the controllers, in general, they are linked by the local area networks or wireless networks. The use of networks brings some advantages to the control systems such as low cost, easy installation and maintenance, and flexible architectures. This results in the drawbacks such as data missing and delays, which are the inherent disadvantages of the networks. Therefore, it is necessary to consider the data missing and delays. The cases occur between the controllers and actuators and between sensors and controllers. The transmission probabilities are α and β. Time delays in control input channel and output measurement channel are τ_1 and τ_2, respectively.

1.1.6 The Characteristics of Switching Signals

In fact, a switching signal plays an important role in a switched delay system, which could rely on the time, the state, itself and an external signal [65, 67]. This leads to more complex dynamic behavior of a switched delay system than that of a switched system or a time delay system. Moreover, such systems still retain the properties of switched systems and time delay systems. When the dynamics of the subsystems are known, the behaviors of switched delay systems are decided by the switching signals. To show this point, we need to introduce an example.

Example 1.1. Consider the following autonomous system composed of two subsystems

$$
\begin{aligned}
x_{k+1} &= A_{\sigma} x_k + A_{d\sigma} x_{k-d}, \\
x_k &= \phi_k,\ k = -d,\ -d+1,\ \cdots,\ 0,
\end{aligned}
$$

where

$$
\begin{aligned}
A_1 &= \begin{bmatrix} 0.4 & 0.05 \\ 0 & 0.35 \end{bmatrix}, \\
A_{d1} &= \begin{bmatrix} 0.025 & 0 \\ -0.1 & -0.35 \end{bmatrix},
\end{aligned}
$$

and

$$
\begin{aligned}
A_2 &= \begin{bmatrix} -0.2 & 0 \\ 0.1 & 0.1 \end{bmatrix}, \\
A_{d2} &= \begin{bmatrix} 0.05 & -0.1 \\ 0 & 0.15 \end{bmatrix}.
\end{aligned}
$$

The state responses of each subsystem are demonstrated in Figs. 1.4 and 1.5. It is seen that two subsystems are unstable.

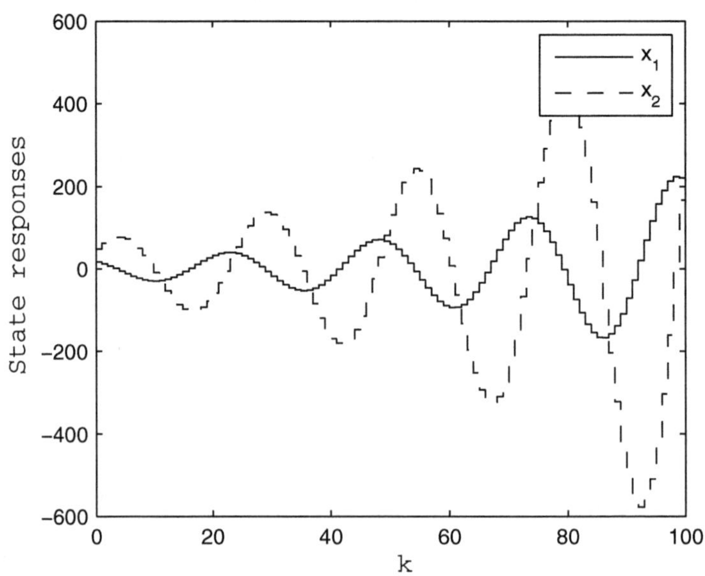

Fig. 1.4 State responses of subsystem 1

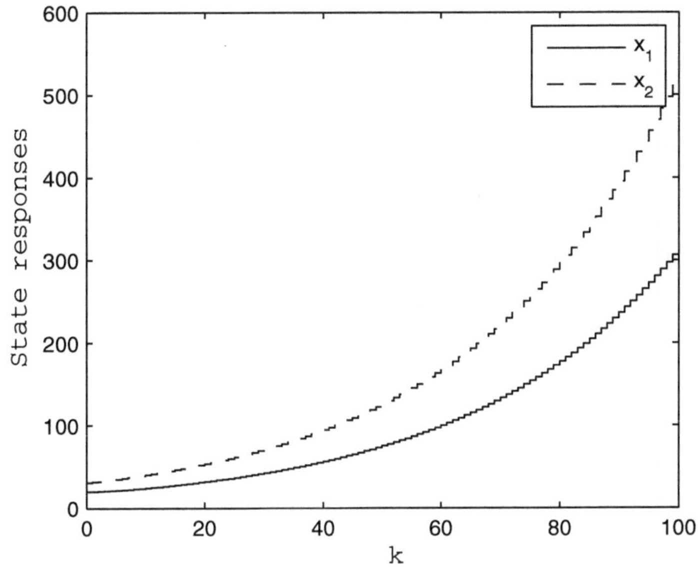

Fig. 1.5 State responses of subsystem 2

1. Time-dependent switching signals. Assume the running time interval is 100 sample time instants. The switching signal $\sigma(t)$ is time-dependent and the dwell time of each subsystem is 10 sample time instants, which is revealed in Fig. 1.6. Under such switching laws, the state responses of switched delay systems are demonstrated in Fig. 1.7.
2. State-dependent switching signals. Assume the running time interval is 100 sample time instants. The switching signal $\sigma(x_2)$ relies on the state x_2. The switching rule is

$$\sigma(x_2) = \begin{cases} 1, & \|x_2\| > 15, \\ 2, & \|x_2\| < 10. \end{cases}$$

   The switching signal is displayed in Fig. 1.8. Under such switching laws, the state responses of switched delay systems are shown in Fig. 1.9.
3. Stochastic switching signals. A random switching signal is to activate a subsystem irregularly but obeys a stochastic distribution. Although all the subsystems are deterministic, a random switching signal renders switched delay systems behave random in nature. For instance, a Markov jump linear system consists in a finite set of linear subsystems and its switches are governed by a finite state Markov chain.
4. Asynchronous switchings. Switched delay systems are generally used to model the systems with large uncertainties or different situations, the controllers or observers are designed for the corresponding subsystems with some performance

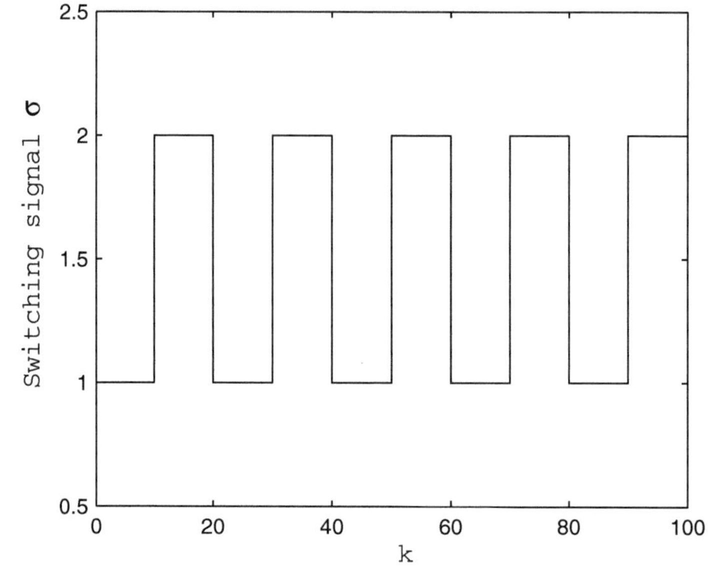

Fig. 1.6 Switching signal with dwell time

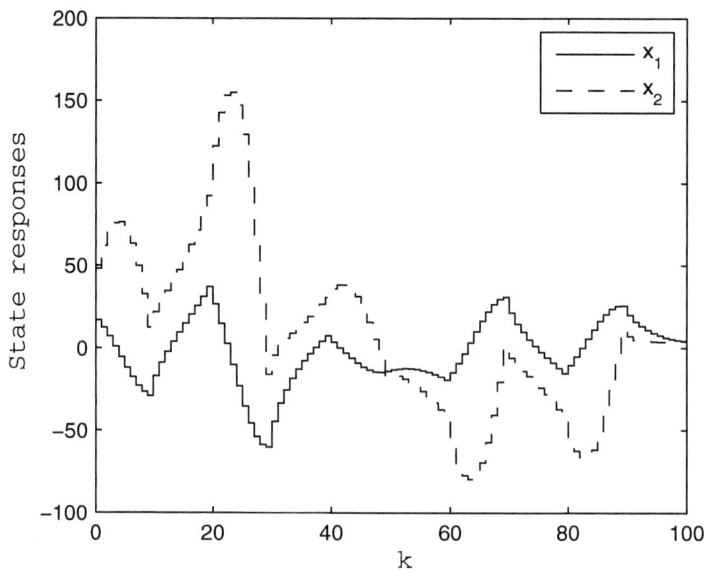

Fig. 1.7 State responses of switched delay systems

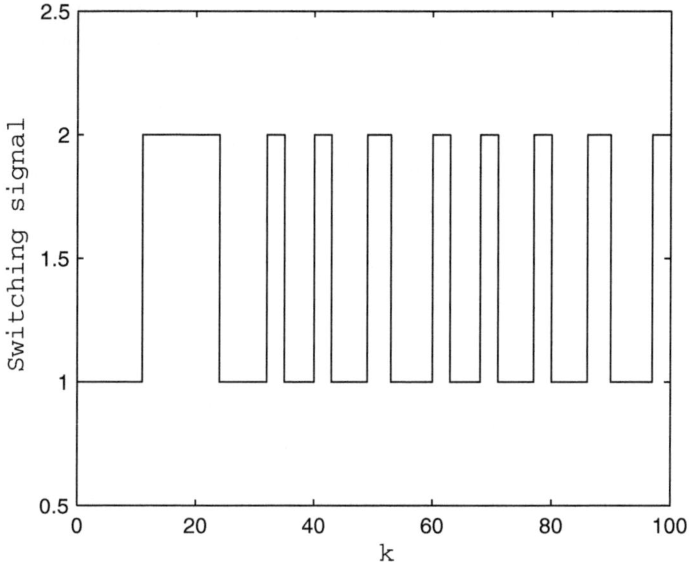

Fig. 1.8 State-dependent switching signal

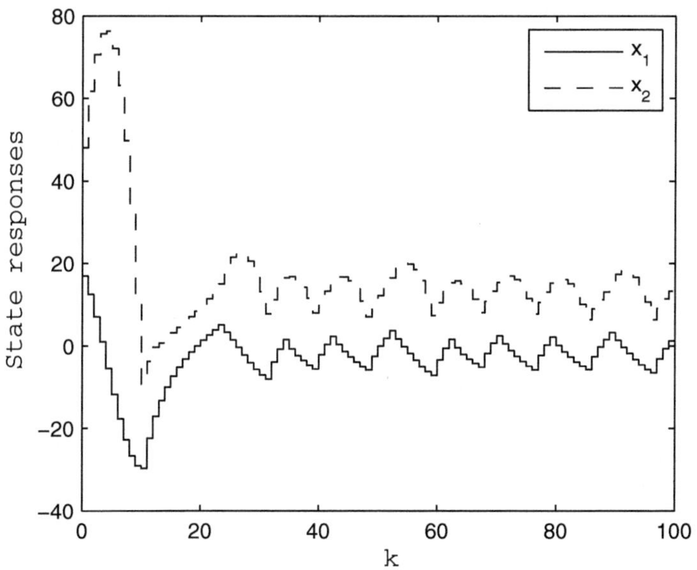

Fig. 1.9 State responses of switched delay systems

index. In the ideal case, the switching of the controllers or observers coincides exactly with that of corresponding subsystems. In real applications, however, since it inevitably takes some time to identify the active subsystem and activate the matched controller or observer, the switching time instant of controllers or observers could lag behind that of practical subsystems, which results in asynchronous switching between the controllers or observers and subsystems. Let us set system (1.3) as an example. For each subsystem, a control input matrix is added as follows:

$$\begin{aligned} x_{k+1} &= A_\sigma x_k + A_{d\sigma} x_{k-d} + B_\sigma u_k, \\ x_k &= \phi_k,\ k = -d,\ -d+1,\ \cdots,\ 0, \end{aligned}$$

where

$$B_1 = \begin{bmatrix} 0.4 \\ 0 \end{bmatrix},\ B_2 = \begin{bmatrix} -0.2 \\ 0.1 \end{bmatrix}.$$

and other matrices are the same with system (1.3). The state feedback controllers

$$\begin{aligned} K_1 &= \begin{bmatrix} -1.2105 & -1.6842 \end{bmatrix}, \\ K_2 &= \begin{bmatrix} 3.8298 & -2.2553 \end{bmatrix}. \end{aligned}$$

are designed to stabilize system (1.3), which is shown in Figs. 1.10 and 1.11.

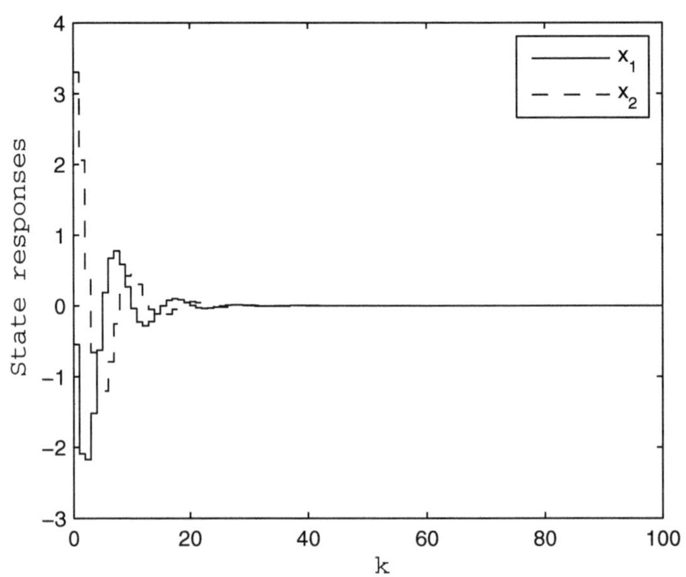

Fig. 1.10 State responses of closed-loop subsystem 1

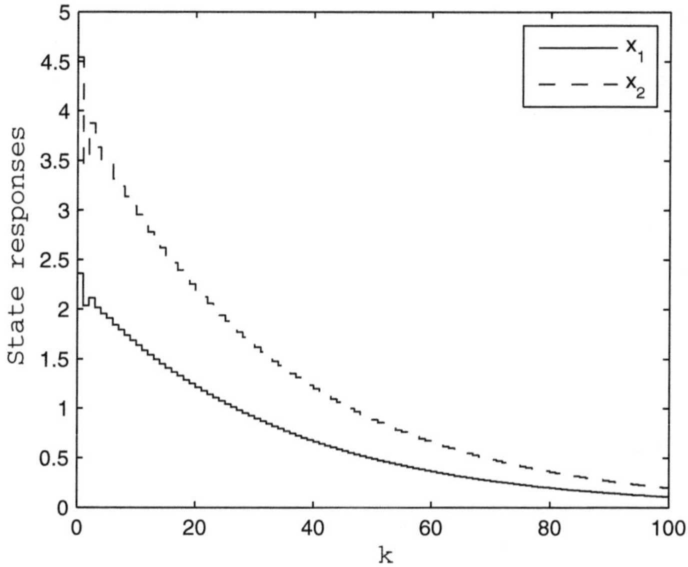

Fig. 1.11 State responses of closed-loop subsystem 2

Assume the running time interval is 100 sample time instants and the dwell time of each subsystem is 10 sample time instants. The switching time instant of controllers lags 5 time instants after that of subsystems. Subsystem 1 and its controller are modeled as mode 1. Subsystem 1 and controller 2 are represented as mode 2. Subsystem 2 and its controller are denoted by mode 3. Subsystem 2 and controller 1 are signified as mode 4. The running time intervals of all the modes are indicated in Fig. 1.12. State responses are displayed in Fig. 1.13, which means that asynchronous switchings lead to destabilize switched delay systems. Hence, the necessities of considering asynchronous switching are obvious. Some primary results under asynchronous switchings for switched systems have been reported [142, 125, 129], to name a few.

It is worth pointing out that although each subsystem is stable, a switching signal could render such system unstable. On the contrary, each subsystem has divergent trajectories, a switching signal could stabilize such system. In switched systems, there exist three identified problems: (1) the stability under arbitrary switching signals; (2) identify certain switching signals for the stability; (3) construct a switching signal to stabilize the systems. These problems are extended to switched delay systems. The developed methods for switched systems have been extended into switched delay systems such as average dwell time and switched Lyapunov function [72, 112, 120, 146, 144]. But some cross terms are generated by the introduction of time delays. More details on the difficulties and challenging will be presented in the subsequent chapters.

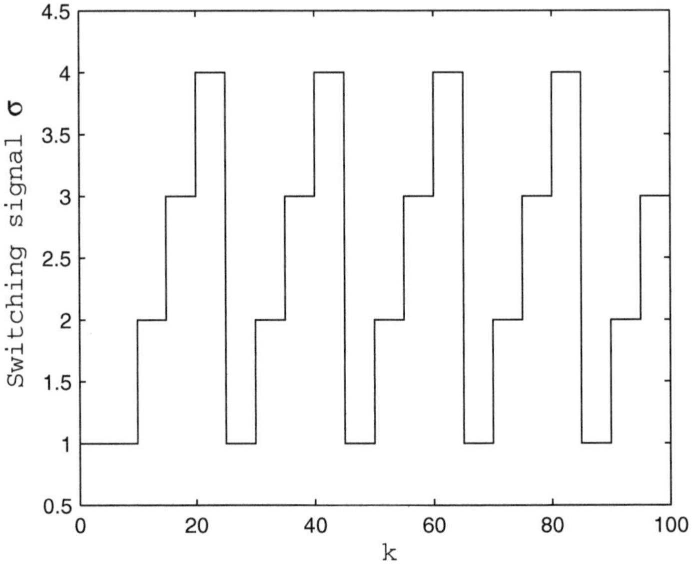

Fig. 1.12 The time intervals of four active modes

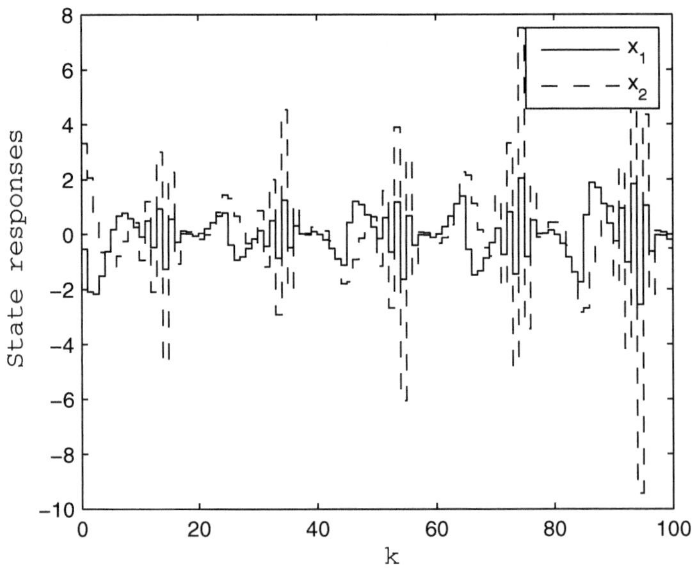

Fig. 1.13 State responses of switched delay systems under asynchronous switchings

1.2 H_∞ Filtering

In real engineering system, since the signals transmitted by the physical devices and measured by the measurement devices are inevitably subject to interference or the influence of the noises, the additive noise signals are received. In order to obtain approximate true signals or the system states, based on a measurement output of the system, various filters are designed depending on different noise signals, called the state estimation [75, 35, 100, 52]. In light of the properties of the noise, it is divided into deterministic and stochastic filtering. The existence of the spectral characteristics of the former allows to design different filters according to the frequency band of the signals such as a low-pass filter, a high-pass filter, a band-pass filter and so on such that the useful signal is passed and the noise signal is suppressed. Such type of filters can be implemented by physical methods or computer algorithms. On the other hand, the noise in the signal could appear stochastically and its characteristics can be described in the statistical sense, which is a stochastic filtering problem. In the 1940s, Wiener proposed Wiener filter designed by means of power spectral decomposition from the perspective of the frequency domain [116]. It requires to solve the the Wiener-Hopf equations. Moreover, the large amount of calculation and a number of storage space render it not applicable for online filtering but only for one dimensional stationary random process. In the 1960s, Kalman proposed a Kalman filter to estimate the true signal mixed with a Gaussian white noise [54, 55]. It is developed in the time domain using the state space description. Since a recursive form is employed to reduce a amount of data stored, such filter has been widely applied. But the disadvantages exist as well. For example, accurate system models are not available for industrial systems. The statistical nature of the noise processes is rarely understood. The continually accumulation and transfer of the rounding and truncation errors cause the missing of symmetric and positive definiteness of error variance matrices. Therefore, a new filter, which can handle modeling errors and uncertainty, is developed, called a robust filter, and has attracted the great attention from scholars and engineers. An effective modification is to model a system with the variational parameters as a interval system and then estimate the states using the interval system theory. A robust H_2 filter for an uncertain system is constructed to tolerate uncertainties in a certain range at the expensive of the bounded estimation error covariance which can be looked up as an extension of the standard Kalman filter [1, 16]. A Kalman filter and a robust H_2 filter are no longer feasible when the statistical properties of the external noise signal are unknown. In 1989, A. Elsayed proposed an H_∞ filter for the energy-bounded noise signal [19]. The main goal lies in minimizing H_∞ norm upper bound of the filtering error system composed of the original system and the filter to ensure that the induced L_2 gain from the noise signal to the estimation error is less than a given index. Take as an example a continuous-time linear time-invariant system:

$$\begin{aligned} \dot{x}(t) &= Ax(t) + Bw(t), \\ y(t) &= Cx(t) + Dw(t), \\ z(t) &= Ex(t), \end{aligned} \tag{1.3}$$

where $x(t)$ is the state, $w(t)$ is the noise signal belonging to L_2 space, $y(t)$ is the measured output and $z(t)$ is an estimated signal. A, B, C, D and E are the constant matrices of the systems. The objective is to design a filter

$$\begin{aligned}\dot{\hat{x}}(t) &= \hat{A}\hat{x}(t) + \hat{B}y(t),\\ \hat{z}(t) &= \hat{C}\hat{x}(t),\end{aligned} \tag{1.4}$$

and to minimize the norm bound γ such that

$$\|z(t) - \hat{z}(t)\|^2 \leq \gamma^2 \|w(t)\|^2, \tag{1.5}$$

where $\hat{x}(t)$ is the state of the filter, $\hat{A}$, $\hat{B}$ and $\hat{C}$ are the constant matrices of the filter and $\hat{z}(t)$ is the output of the filter.

The H_∞ filter has two advantages. One is the process and measurement noises can be any energy-bounded signal instead of the Gaussian signal. The other is a filter is robust to the uncertainty in the model such as parameter and structure uncertainties. Some H_∞ filter design methods have been developed such as polynomial equations [36], interpolation [95], algebraic Riccati equation (ARE) [32] and linear matrix inequality (LMI) method [62]. Polynomial equations and interpolation use the transfer functions, which leads to the complexity of formula, especially in the multi-variable cases. On the contrast, ARE, a state space method, is more popular than the frequency domain method because the solution is simple. But it often requires to determine in advance unknown parameters, the choice of which affects not only the performance of the solution but also the solvability of the problem. Moreover, the solution of a Riccati equation is provided by iterative methods, the convergence of which can not be guaranteed. In the 1990s, with the development of an interior point method for solving a convex optimization problem, linear matrix inequality method is developed to overcome the shortcomings of the Riccati equation approach [30, 24, 34].

On the other hand, many practical systems are originally modeled as high-order models because it is straightforward to obtain the model formulation using many variables. However, this causes the great difficulties in analysis and synthesis of the systems. Thus, it is desirable to replace these high-order models with reduced-order ones for reducing the computational complexities in some given criteria without incurring much loss of performance or information, which has motivated the study of the model reduction problem. During the past few years, many criteria have been proposed including the L_2 model reduction [102, 108] and the H_2 model reduction [134, 38]. Recently, the problem of H_∞ model reduction is addressed by making use of LMIs technique for singular systems [130], time-delay system [29, 121], Markovian jump systems [143] and switched systems [145]. It is worth mentioning that the criteria for the discrete-time context in [130, 29, 143] are formulated in terms of LMIs with some non-convex conditions, which make LMIs difficult to find the numerical solution. Although many numerical approaches have been proposed to overcome this difficulty, for example, the cone complementarity linearization method

[18] and sequential linear programming matrix method [61], the computation is very heavy. However, the strict LMI formulation is desired to avoid it.

1.3 Fault Detection

Whether the normal operation of the system or not directly affects the economic interests and even is related to the safety of people's life. If the failure of the system can not be detected in time, it may cause the paralysis of the entire system and the huge losses of persons and production, and even lead to disastrous consequences. For example, the former Soviet Union, the 'Chernobyl' nuclear power plant explosion caused more than 30 deaths and 320 million persons suffering from nuclear radiation and contaminated 15 million square kilometers of land. European 'Ariane' carrier rocket crashed due to the failure of a navigation system, which results in the huge losses of hundreds of millions of dollars. U.S. carrier rocket 'Hercules', 'Athena', 'Delta' etc. launch abortively a total of five times causing a direct economic loss of over 30 billion U.S. dollars. The wing fire of French 'Concorde' airliner lead to explode during the process of takeoff killing all 109 persons on board and 5 persons on the ground. Therefore, there is an urgent need to establish a system to monitor the operation of the system state, constantly test the system state changes and detect the system faults, and then take measures to prevent the system accidents. Fault detection and diagnosis came into being and rapid developed, which has opened up a new way to improve the reliability and security of the system. It is attached the importance in the control and system community since it plays an crucial role in whether industrial, military, aviation, aerospace or other fields [9, 48, 47].

A fault of the system refers to the abnormal operation state of the system and can lead to system malfunctions. Fault detection and diagnosis (FDD) refers to in the operation environment testify the equipment running status information to identify the occurrence, the positions and the components of the fault and make the necessary decisions. FDD is originally derived from a hardware redundancy method. Redundant components are used to provide the information to determine the faulty components, but this method greatly increases the cost and space of the systems. In 1971, Beard determined the system fault by comparing the output of observers [4]. It marks the beginning of analytical redundancy FDD that is low cost and easy to implement. In 1976, Willsky published the first review paper relevant to fault detection [119]. In 1990, Frank presented a robust fault detection method based on unknown input observer [21]. Isermann summarized the fault detection method based on parameter estimation [47]. There are three major methods: the mathematical model, signal processing and knowledge-based methods [49, 23]. The first method is to generate the residuals based on the system mathematical model and then analyze the residuals to perform FDD. In the industrial system, the disturbance inputs (modeling errors, the system noise and interference signals) and fault signals often exist at the same time and mutually couple. The disturbance input is a source of error alarms. Hence, it is required the fault detection system is sensitive to the fault signal and robust to the disturbance signal. The unknown input observer-based method is proposed to consider disturbances as the unknown input and then decouple it from the estimated states. An eigen-structure assignment-based method is to assign the observer's left eigenvectors orthogonal to interference direction matrices so that the residual is robust to the interference signal. These methods mentioned above can only deal with the uncertainties of a special structure. Since the H_∞ performance optimization method is applicable to handle the general uncertainties, it is applied to the fault detection system in recent years [14, 151, 152, 39, 147].

1.4 Organization of the Book

This book contains two main parts: robust filtering and fault detection. The former is involved with 2-5 chapters while the latter 6-7 chapters.

Chapter 1 contains the development and motivation of switched delay systems, essential concepts for robust filtering and fault detection.

In Chapter 2, the problem of filter design for discrete-time switched systems with state delays is investigated. Based on a novel linearization technique, delay-independent and sufficient conditions for the existence of filters are obtained, by solving which a desired filter can be constructed.

In Chapter 3, the problem of delay-dependent H_∞ filter design for a class of discrete-time switched systems with state delays is studied. By using switched Lyapunov function method and free-weighting matrix method, choosing a new Lyapunov function and utilizing linearization technique, delay-dependent sufficient conditions on the existence of a desired filter are presented.

In Chapter 4, the problem of delay-dependent exponential H_∞ filtering for discrete-time switched delay systems is investigated under average dwell time (ADT) switching signals. By introducing a proper factor to construct a novel Lyapunov function and using ADT approach, sufficient conditions for the solvability of this problem are obtained. Furthermore, the problem of delay-dependent exponential H_∞ model reduction is also studied for such kind of switched delay systems. Sufficient conditions on the existence of the reduced-order model are given.

In Chapter 5, the problem of exponential H_∞ filtering for a class of continuous-time switched system with interval time-varying delays is studied. Based on the free weighting matrix approach and average dwell time technology, delay-dependent sufficient conditions for the existence of such filter are derived. Furthermore, the problem of delay-dependent exponential H_∞ model reduction is also studied for such kind of switched delay systems. Sufficient conditions on the existence of the reduced-order model and parameter design are presented.

In Chapter 6, the problem of robust fault detection for discrete-time switched systems with state delays under an arbitrary switching signal is studied. The fault detection filter is used as the residual generator. Our aim is to design the robust fault detection filter such that, for unknown inputs, control inputs, and model uncertainties, the estimation error between the residuals and faults is minimized. Utilizing H_∞ theory, the problem of robust fault detection is converted into an H_∞ filtering problem. By a switched Lyapunov functional approach, a sufficient condition for the solvability of this problem is established.

In Chapter 7, The robust fault detection problem for continuous-time switched systems with state delays is addressed. Our objective is to design the fault detection filter as the residual generator in order to minimize the error between the residuals and the faults in the presence of the modeling errors, the unknown inputs and the control one. The addressed problem is converted into an auxiliary H_∞ filtering problem. By constructing a new Lyapunov-Krasovskii functional and using average dwell time approach, a sufficient condition for the solvability of this problem is established.

In Chapter 8, the perspectives and future extensions of switched delay systems are provided.

1.5 Notations

The notations throughout this book are fairly standard. Matrices or vectors, if their dimensions are not explicitly stated, are assumed to be compatible for algebraic operations. The list of Symbols are the following:

$s.t.$	such that
R^n	the set of all n-dimensional real vectors
I	an identity matrix
0	a zero matrix
x^T	the transpose of a vectorx
A^T	the transpose of a matrix A
$diag\{\cdots\}$	a block-diagonal matrix
A^{-1}	the inverse of a matrix A
$\lambda(A)$	the set of eigenvalues of a matrix A
$\lambda_m(A)$	the minimum eigenvalues of a matrix A
$\lambda_M(A)$	the maximum eigenvalues of a matrix A
$P<0$	a real symmetric and negative definite matrix
$P\leq 0$	a real symmetric and negative semidefinite matrix
$P>0$	a real symmetric and positive definite matrix
$P\geq 0$	a real symmetric and positive semidefinite matrix
$\|A\|_p$	the induced l_p norm of a matrix A
$\|x\|_p$	the l_p norm of a vector x

In addition, $*$ denotes the symmetric terms in symmetric block matrices.

Part I

H_∞ Filter Design

2

Delay-Independent H_∞ Filtering for Switched Systems with a Constant Delay

This chapter deals with the problem of H_∞ filtering for discrete-time switched linear system with a constant delay. Switched linear filters are designed to ensure that the filtering error systems under arbitrary switching signals are stable with a prescribed H_∞ noise attenuation level. Switched Lyapunov function approach is employed to analyze the stability of the filtering error systems. A new linearization technique is proposed to construct the parameters of filters. Simulation results verify the effectiveness of the developed method.

2.1 Introduction

Considerable effort has been devoted to the Kalman filtering during the past few decades. When a priori information on the external noises such as its statistics is not precisely known, the celebrated Kalman filtering scheme is no longer applicable. In this cases, H_∞ filtering method is introduced where the noise signals are assumed to be arbitrary signals with energy bounded, and the aim is to minimize the l_2 gain from the noise signals to the estimation errors. On another front line, switched systems have received much attention mainly because they have strong engineering background such as Networked Control Systems (NCSs), power systems, automotive engine control, robotic manufacture and so on (see, also, [106, 107, 70] and the references therein). In [106], the fundamental issues for switched linear systems are focused on such as controllability, observability, feedback stabilization, optimization and periodic switching. The problems of NCSs' stability and disturbance attenuation are formulated as the corresponding problems of switched systems [70]. The existence of a common Lyapunov function for all subsystems means that a switched system is asymptotically stable under an arbitrary switching law [67]. However, in most cases, a common Lyapunov function for all subsystems is unknown or do not exist. To overcome the difficulty, multiple Lyapunov functions (MLF) is interesting to analyze the stability for switched systems [6]. The issues of stability, L_2-gain analysis and H_∞ control for switched systems are addressed by proposing general Lyapunov-like functions based on multiple Lyapunov functions method [148]. For discrete-time

D. Wang et al.: *Robust Filtering & Fault Detect. of Switched Delay Systems*, LNCIS 445, pp. 23–33.
DOI: 10.1007/978-3-642-37685-6_2

switched systems, a switched Lyapunov function approach as a kind of MLF is proposed to obtain arbitrary switching signals based on the characteristics of discrete time instants to get rid of the shortcoming of common Lyapunov function [10]. Using such method, the issues of stability, stabilization and filtering etc. have been studied. With the high requirements of the precessing speed and precision in industrial production, time delay systems have become one of hot topics. In particular, switched delay systems have attracted great attention from the scholars all over the world. The stability analysis and synthesis issues in switched delay systems have been addressed such as [111, 112, 44, 15, 58, 144]. On the other hand, state estimation and filtering of dynamic systems is a subject of great practical and theoretical significance. When the celebrated Kalman filtering scheme is no longer applicable in the case that a priori information on the external noises is not precisely known, an alternatives, the H_∞ filtering scheme is proposed for switched systems [91, 46, 146], [15], to name a few. In [91], sufficient conditions for robust H_2 and H_∞ performance index for discrete-time switched linear systems are given. A robust H_∞ switched linear filter for switched linear discrete-time systems with polytopic uncertainties is designed based on the SLF approach in [146]. The existence condition of such filter is derived by introduction of slack variables to eliminate the cross coupling of systems matrices and Lyapunov matrices among different subsystems. It should be pointed out that in all the papers mentioned above, time delays are not taken into account. It is well known that a great number of practical systems are unavoidably subject to delays, which usually degrade the performance of the systems. Hence, research efforts have been focused on the problem of H_∞ filtering design for switched systems with time delays, see for example, [15] and the references therein. An H_∞ filtering design is developed by using switched Lyapunov function approach for discrete-time switched systems with a time delay in [15]. An extension of the result in [15] is provided by introducing a new linearized technique to decouple the Lyapunov and the system matrices.

The chapter is organized as follows. In Section 2.2, the problem under consideration is formulated with some basic concepts and lemmas. Section 2.3 presents some lemmas for performance analysis. In Section 2.4, the filter is designed by a new linearized technique. A numerical example is provided to show the effectiveness of the results in Section 2.5, and then, Section 2.6 concludes this chapter.

2.2 Problem Formulation

Consider the following discrete-time switched systems with a constant delay:

$$\begin{aligned} x_{k+1} &= A_\sigma x_k + A_{d\sigma} x_{k-d} + B_\sigma w_k, \\ y_k &= C_\sigma x_k + C_{d\sigma} x_{k-d} + D_\sigma w_k, \\ z_k &= G_\sigma x_k, \\ x_l &= \phi_l,\, l = -d, -d+1, \ldots, 0, \end{aligned} \tag{2.1}$$

where $x_k \in R^n$ is the state, $y_k \in R^r$ is the measured output, $z_k \in R^q$ is the signal to be estimated. $w_k \in R^p$ is the disturbance input, which is assumed to belong to $l_2[0,\infty)$. σ denoting for simplicity $\sigma(k) : [0, \infty) \to \mathscr{N}$ is the switching signal, specifying which subsystem is activated at the switching instant. $\mathscr{N}$ and d denote an index set of subsystems and a known state delay, respectively. The matrices A_p, A_{dp}, B_p, C_p, C_{dp}, D_p and $G_p, p \in \mathscr{N}$ of the pth subsystem are of the appropriate dimensions.

In this chapter, we are interested in designing a full-order switched filter described by

$$\begin{aligned} \hat{x}_{k+1} &= A_{f\sigma}\hat{x}_k + B_{f\sigma}y_k, \\ \hat{z}_k &= C_{f\sigma}\hat{x}_k, \end{aligned} \tag{2.2}$$

where $\hat{x}_k \in R^n$ and $\hat{z}_k \in R^q$. σ is defined in (2.1). The matrices A_{fp}, B_{fp} and C_{fp} are the parameters of the pth subsystem of filter (2.2) to be determined.

Remark 2.1. The linear filters for discrete-time switched systems are designed in [15], where the switching signal is stochastically selected in a finite set. Our objective is to design a filter the switching signal of which is the same with that of system (2.1). Thus, when the switching signal changes, the switchings between the subsystems in system (2.1) and filter (2.2) occur simultaneously. Additionally, the switching signals are assumed to be unknown a priori, but their instantaneous values are available in real-time implementation.

Here, in view of (2.1) and (2.2), the filtering error systems are described by

$$\begin{aligned} \tilde{x}_{k+1} &= \tilde{A}_\sigma\tilde{x}_k + \tilde{A}_{d\sigma}E\tilde{x}_{k-d} + \tilde{B}_\sigma w_k, \\ \tilde{z}_k &= \tilde{C}_\sigma\tilde{x}_k, \end{aligned} \tag{2.3}$$

where

$$\begin{aligned} \tilde{x}_k^T &= \begin{bmatrix} x_k^T & \hat{x}_k^T \end{bmatrix}, \\ \tilde{z}_k &= z_k - \hat{z}_k,\ E = \begin{bmatrix} I & 0 \end{bmatrix}, \\ \tilde{A}_\sigma &= \begin{bmatrix} A_\sigma & 0 \\ B_{f\sigma}C_\sigma & A_{f\sigma} \end{bmatrix}, \\ \tilde{A}_{d\sigma} &= \begin{bmatrix} A_{d\sigma} \\ B_{f\sigma}C_{d\sigma} \end{bmatrix}, \\ \tilde{B}_\sigma &= \begin{bmatrix} B_\sigma \\ B_{f\sigma}D_\sigma \end{bmatrix}, \\ C_\sigma &= \begin{bmatrix} G_\sigma - D_{f\sigma}C_\sigma & -C_{f\sigma} \end{bmatrix}. \end{aligned}$$

The problem addressed in this paper is to design an H_∞ filter in the form of (2.2) such that the following specifications are met for filtering error system (2.3) under arbitrary switching signals:

(1) Filtering error system (2.3) is asymptotically stable when $w_k = 0$;
(2) Filtering error system (2.3) guarantees, under zero-initial condition, $\|\tilde{z}_k\|_2 \leq \gamma\|w_k\|_2$ for all nonzero $w_k \in l_2[0,\infty)$ and a given positive constant γ.

The following lemma is a powerful tool and exploited to linearize some nonlinear inequalities to linear inequalities.

Lemma 2.2. *(Schur complement) [5] Given a symmetric and partitioned matrix*

$$E(X) = \begin{bmatrix} A(X) & C(X) \\ C^T(X) & B(X) \end{bmatrix},$$

where $A(X)$, $B(X)$ and $C(X)$ are the matrix functions of an unknown variable X with appropriated dimensions. In addition, $A(X)$ and $B(X)$ are square and nonsingular. Then, $E(X) < 0$ if and only if either of the following conditions is true

$$\begin{cases} A(X) < 0 \\ B(X) - C^T(X)A^{-1}(X)C(X) < 0 \end{cases}, \tag{2.4}$$

and

$$\begin{cases} B(X) < 0 \\ A(X) - C(X)B^{-1}(X)C^T(X) < 0 \end{cases}. \tag{2.5}$$

If $A(X)$ is nonsingular, then, $B(X) - C^T(X)A^{-1}(X)C(X) < 0$ is the schur complement of $A(X)$ in $E(X)$. Similarly, if $B(X)$ is nonsingular, $A(X) - C(X)B^{-1}(X)C^T(X) < 0$ is the schur complement of $B(X)$ in $E(X)$. The second inequalities in (2.4) and (2.5) are nonlinear constraints in X. Using this lemma, it follows that nonlinear matrix inequalities of the form (2.4) or (2.5) can be converted to linear matrix inequalities which are solvable using LMI toolbox in Matlab.

2.3 Stability and H_∞ Performance Analysis

In this section, the following Lemma provides an H_∞ performance analysis for filtering error system (2.3), which will play an important role in our filter design.

Lemma 2.3. *For a given scalar $\gamma > 0$ and the positive integer d, the filtering error system (2.3) under arbitrary switching signals is stable with H_∞ norm bound γ if there exist symmetric and positive definite matrices P_p and Q_p, $p \in \mathcal{N}$ such that the following inequality holds,*

$$\begin{bmatrix} -P_j^{-1} & \tilde{A}_i & \tilde{A}_{di} & \tilde{B}_i & 0 \\ * & -P_i + Q_i & 0 & 0 & \tilde{C}_i^T \\ * & * & -Q_l & 0 & 0 \\ * & * & * & -\gamma^2 I & 0 \\ * & * & * & * & -I \end{bmatrix} < 0, \tag{2.6}$$

where $\{i,\ j,\ l\} \in \mathcal{N}$.

Proof. The proof is divided into two parts: stability analysis and performance index testification of the filtering error systems. For filtering error systems (2.3), the following Lyapunov function is constructed

$$V_{\sigma k} = \tilde{x}_k^T P_\sigma \tilde{x}_k + \sum_{h=k-d}^{k-1} \tilde{x}_h^T Q_\sigma \tilde{x}_h, \tag{2.7}$$

where P_σ and Q_σ are symmetric and positive definite matrices.

Taking the forward difference $\Delta V_{\sigma k} = V_{\sigma k+1} - V_{\sigma k}$ along the trajectory of system (2.3) yields

$$\Delta V_{\sigma k} = \tilde{x}_{k+1}^T P_j \tilde{x}_{k+1} - \tilde{x}_k^T P_i \tilde{x}_k + \tilde{x}_k^T Q_i \tilde{x}_k - \tilde{x}_{k-d}^T Q_l \tilde{x}_{k-d}, \tag{2.8}$$

where $j = \sigma(k+1)$, $i = \sigma(k)$ and $l = \sigma(k-d)$.

First of all, for $w_k \equiv 0$, we have

$$\begin{aligned} \Delta V_{\sigma k} &= \xi_k^T \left(\begin{bmatrix} \tilde{A}_i^T \\ \tilde{A}_{di}^T \end{bmatrix} P_j \begin{bmatrix} \tilde{A}_i & \tilde{A}_{di} \end{bmatrix} \right) \xi_k, \\ &+ \xi_k^T \left(\begin{bmatrix} -\tilde{P}_i + Q_i & 0 \\ 0 & -Q_l \end{bmatrix} \right) \xi_k, \end{aligned} \tag{2.9}$$

where $\xi_k = \begin{bmatrix} \tilde{x}_k^T & \tilde{x}_{k-d}^T \end{bmatrix}^T$.

Using Schur complement, it is obtained from (2.6) that

$$\Delta V_{\sigma k} < -\rho \left\| \xi_k \right\|^2$$

for a sufficiently small $\rho > 0$ and $\xi_k \neq 0$, which establishes the asymptotical stability of system (2.3) with $w_k = 0$.

Secondly, for any nonzero $w_k \in l_2 [0, \infty)$, we consider the following performance index:

$$J = \sum_{k=0}^{K-1} \left(\tilde{z}_k^T \tilde{z}_k - \gamma^2 w_k^T w_k \right)$$

where K is an arbitrary positive integer.

Under the zero-initial condition $\tilde{x}_0 = 0$, one has

$$\begin{aligned} J &= \sum_{k=0}^{K-1} \left(\tilde{z}_k^T \tilde{z}_k - \gamma^2 w_k^T w_k + \Delta V_{\sigma k} \right) - V_{\sigma K} \\ &\leq \sum_{k=0}^{K-1} \left(\tilde{z}_k^T \tilde{z}_k - \gamma^2 w_k^T w_k + \Delta V_{\sigma k} \right) \\ &= \tilde{\xi}_k^T \left(\begin{bmatrix} \tilde{A}_i^T \\ \tilde{A}_{di}^T \\ \tilde{B}_i^T \end{bmatrix} P_j \begin{bmatrix} \tilde{A}_i & \tilde{A}_{di} & \tilde{B}_i \end{bmatrix} \right) \tilde{\xi}_k \\ &+ \tilde{\xi}_k^T \left(\begin{bmatrix} -P_i + Q_i + \tilde{C}_i^T \tilde{C}_i & 0 & 0 \\ 0 & -Q_l & 0 \\ 0 & 0 & -\gamma^2 I \end{bmatrix} \right) \tilde{\xi}_k \end{aligned}$$

where $\tilde{\xi}_k = \begin{bmatrix} \tilde{x}_k^T & \tilde{x}_{k-d}^T & w_k^T \end{bmatrix}^T$.

It follows from (2.6) and Schur complement that $J < 0$. Then, for any nonzero $w_k \in l_2[0,\infty)$, we have $\|\tilde{z}_k\|_2^2 < \gamma^2 \|w_k\|_2^2$, which completes the proof.

Remark 2.4. If no time delay exists in system (2.1), namely, $d = 0$, which is considered in [10], switched Lyapunov function method means that $i = j$ shows the occurrence of no switching while $i \neq j$ represents the switching from the *ith* subsystem to the *jth* subsystem. The existence of a time delay d leads to the appearance of the *lth* subsystem. It is allowed to have $i \neq j \neq l$. Therefore, three subsystems may be involved in condition (2.6).

Remark 2.5. Note that a time delay d does not appear in the condition (2.6), which means that the condition is applicable to arbitrary size of a constant time delay. This is a delay-independent result. From (2.6), it is also noted that there exist the inverse of matrix P_p, which renders matrix inequalities (2.6) nonlinear.

Motivated by the idea in [15], the following Lemma is given based on Lemma 2.3 by introducing a new matrix to eliminate the product resulting from P_j^{-1}.

Lemma 2.6. *For a given scalar $\gamma > 0$, condition (2.6) hold if there exist symmetric and positive definite matrices R_p, Φ_p and Ω, $p \in \mathcal{N}$ such that the following inequality holds,*

$$\begin{bmatrix} -R_j & \tilde{A}_i\Omega & \tilde{A}_{di}\Omega & \tilde{B}_i & 0 & 0 \\ * & R_i - (\Omega+\Omega^T) & 0 & 0 & \Omega^T\tilde{C}_i^T & \Omega^T \\ * & * & \Psi_l - (\Omega+\Omega^T) & 0 & 0 & 0 \\ * & * & * & -\gamma^2 I & 0 & 0 \\ * & * & * & * & -I & 0 \\ * & * & * & * & * & -\Psi_i \end{bmatrix} < 0. \quad (2.10)$$

Proof. For symmetric and positive definite matrices R_p and a matrix Ω, we have

$$(R_q - \Omega) P_q^{-1} (R_q - \Omega)^T \geq 0,$$

and further

$$-\Omega R_q^{-1} \Omega^T \leq R_q - \left(\Omega + \Omega^T\right). \quad (2.11)$$

Similarly, one has

$$-\Omega \Phi^{-1} \Omega^T \leq \Phi_4 - \left(\Omega + \Omega^T\right). \quad (2.12)$$

From (2.11) and (2.12), it is observed that (2.10) means that

$$\begin{bmatrix} -R_j & \tilde{A}_i\Omega & \tilde{A}_{di}\Omega & \tilde{B}_i & 0 & 0 \\ * & -\Omega R_q^{-1}\Omega^T & 0 & 0 & \Omega^T\tilde{C}_i^T & \Omega^T \\ * & * & -\Omega\Phi^{-1}\Omega^T & 0 & 0 & 0 \\ * & * & * & -\gamma^2 I & 0 & 0 \\ * & * & * & * & -I & 0 \\ * & * & * & * & * & -\Psi_i \end{bmatrix} < 0. \quad (2.13)$$

Replacing R_p and Φ_p with P_i^{-1} and Q_i^{-1} and performing a congruence transformation to (2.10) via $diag\left\{I,\ \Omega^{-T},\ \Omega^{-T},\ I,\ I,\ I\right\}$ yields (2.6). This proof is completed.

2.4 H_∞ Filter Design

In this section, based on the result in [15], an H_∞ filter for discrete-time switched systems with state delays is designed in the form of (2.2). The parameters of such filter are constructed by the following theorem.

Theorem 2.7. *Consider system (2.1), for a given scalar $\gamma > 0$, if there exist symmetric and positive definite matrices R_{1i}, R_{3i}, X_{1i}, X_{3i}, and matrices R_{2i}, X_{2i}, Z_i, Y_i, H_i, L_i, M_i, S_i, $\forall\{i,\ j,\ l\} \in \{1,2,\ldots,N\}$ such that the following inequality holds,*

$$\begin{bmatrix} -\Psi_{11} & \Psi_{12} & \Psi_{13} & \Psi_{14} & 0 & 0 \\ * & \tilde{\Psi}_{22} & 0 & 0 & \Psi_{25} & \Psi_{26}^T \\ * & * & \tilde{\Psi}_{33} & 0 & 0 & 0 \\ * & * & * & -\gamma^2 I & 0 & 0 \\ * & * & * & * & -I & 0 \\ * & * & * & * & * & -\Psi_{66} \end{bmatrix} < 0, \tag{2.14}$$

where

$$\begin{aligned}
\Psi_{11} &= \begin{bmatrix} R_{1j} & R_{2j} \\ * & R_{3j} \end{bmatrix}, \\
\Psi_{12} &= \begin{bmatrix} Z_iA_i & Z_iA_i \\ Y_iA_i + H_iC_i + L_i & Y_iA_i + H_iC_i \end{bmatrix}, \\
\Psi_{13} &= \begin{bmatrix} Z_iA_{di} & Z_iA_{di} \\ Y_iA_{di} + H_iC_{di} & Y_iA_{di} + H_iC_{di} \end{bmatrix}, \\
\Psi_{14} &= \begin{bmatrix} Z_iB_i \\ Y_iB_i + H_iD_i \end{bmatrix}, \\
\Psi_{22} &= \begin{bmatrix} R_{1i} & R_{2i} \\ * & R_{3i} \end{bmatrix}, \\
\Psi_{25} &= \begin{bmatrix} G_i^T - S_i^T \\ G_i^T \end{bmatrix}, \\
\Psi_{26} &= \begin{bmatrix} Z_i & Z_i \\ Y_i + M_i & Y_i \end{bmatrix}, \\
\Psi_{33} &= \begin{bmatrix} X_{1l} & X_{2l} \\ * & X_{3l} \end{bmatrix}, \\
\Psi_{66} &= \begin{bmatrix} X_{1i} & X_{2i} \\ * & X_{3i} \end{bmatrix}, \\
\tilde{\Psi}_{22} &= \Psi_{22} - \Psi_{26} - \Psi_{26}^T, \\
\tilde{\Psi}_{33} &= \Psi_{33} - \Psi_{26} - \Psi_{26}^T.
\end{aligned}$$

Then, there exists a filter in the form of (2.2) such that system (2.3) under arbitrary switching signals is asymptotically stable with H_∞ norm bound γ. Moreover, if LMI (2.14) has a feasible solution, then the parameters of the admissible filter can be constructed by

$$\begin{bmatrix} A_{fi} & B_{fi} \\ C_{fi} & 0 \end{bmatrix} = \begin{bmatrix} V_i^{-1} L_i M_i^{-1} V_i & V_i^{-1} H_i \\ S_i M_i^{-1} V_i & 0 \end{bmatrix}, \tag{2.15}$$

where $V_i \in \mathscr{R}^{n\times n}$ is any invertible matrix (for example, V_i could be set as I).

Proof. Note that from (2.14), we have

$$\begin{bmatrix} Z_i + Z_i^T & Z_i + Y_i^T + M_i^T \\ * & Y_i + Y_i^T \end{bmatrix} > \begin{bmatrix} R_{1i} & R_{2i} \\ * & R_{3i} \end{bmatrix} > 0, \tag{2.16}$$

which means that Z_i and Y_i are nonsingular.

Premultiplying $\begin{bmatrix} I, & -I \end{bmatrix}$ and postmultiplying $\begin{bmatrix} I, & -I \end{bmatrix}^T$ to (2.16), one has $-M_i - M_i^T > 0$, which implies that M_i is nonsingular. And thus, if (2.14) holds, there exist nonsingular matrices V_i and U_i satisfying $M_i = V_i U_i$.

Now, introducing the following matrix variables and functions:

$$\begin{aligned} F_i^T &= \begin{bmatrix} Z_i & 0 \\ Y_i & V_i \end{bmatrix}, \\ \Omega F_i &= \begin{bmatrix} I & I \\ U_i & 0 \end{bmatrix}, \\ H_i &= V_i B_{fi},\ L_i = V_i A_{fi} U_i, \\ S_i &= C_{fi} U_i,\ M_i = V_i U_i, \\ R_j &= F_i^{-T} \Psi_{11} F_i^{-1}, \\ R_i &= F_i^{-T} \Psi_{22} F_i^{-1}, \\ \Phi_l &= F_i^{-T} \Psi_{33} F_i^{-1}, \\ \Phi_i &= F_i^{-T} \Psi_{66} F_i^{-1}. \end{aligned} \tag{2.17}$$

By using (2.3) and (2.17), we can get

$$\begin{aligned} F_i^T \tilde{A}_i \Omega F_i &= \Psi_{12}, \\ F_i^T \tilde{A}_{di} \Omega F_i &= \Psi_{13}, \\ F_i^T \tilde{B}_i &= \Psi_{14}, \\ \tilde{C}_i \Omega F_1 &= \Psi_{25}, \\ F_i^T \Omega F_i &= \Psi_{26}. \end{aligned} \tag{2.18}$$

Premultiplying $diag\{F_i^T, F_i^T, F_i^T, I, I, F_i^T\}$ and postmultiplying $diag\{F_i, F_i, F_i, I, I, F_i\}$ to (2.10) yields (2.14) by using (2.17) and (2.18). Moreover, it is easy to see from (2.17) that the parameters of an admissible filter are given by (2.15). The proof is completed.

Remark 2.8. In this note, by introducing F_i^T and ΩF_i, we get

$$
\begin{aligned}
&F_i^T \tilde{A}_{di} \Omega F_i \\
&= \begin{bmatrix} Z_i A_{di} & Z_i A_{di} \\ Y_i A_{di} + V_i B_{fi} C_{di} & Y_i A_{di} + V_i B_{fi} C_{di} \end{bmatrix} \\
&= \Psi_{13}
\end{aligned}
$$

where $V_i B_{fi} C_{di}$ is nonlinear for V_i and B_{fi}. Moreover, $Y_i A_{di} + V_i B_{fi} C_{di}$ can not be obtained by solving (2.14). Thus, it is correct to define $H_i = V_i B_{fi}$. This yields that $Y_i A_{di} + H_i C_{di}$ is linear for Y_i and H_i. Then, the desired filter can be constructed by Theorem 2.7.

2.5 An Illustrative Example

In this section, an example is presented to check the validity of the proposed filtering design algorithms.

Example 2.9. Consider system (2.1) composed of two subsystems with the following parameters:

$$
\begin{aligned}
A_1 &= \begin{bmatrix} 0.4 & 0.05 \\ 0 & 0.35 \end{bmatrix}, \\
A_{d1} &= \begin{bmatrix} 0.025 & 0 \\ -0.1 & -0.35 \end{bmatrix}, \\
B_1 &= \begin{bmatrix} 0.34 & -0.3 \end{bmatrix}^T, \\
C_1 &= \begin{bmatrix} 0.29 & 0.15 \end{bmatrix}, \\
C_{d1} &= \begin{bmatrix} 0.02 & 0 \end{bmatrix}, \\
D_1 &= 0.02, \\
G_1 &= \begin{bmatrix} 0.24 & 0.23 \end{bmatrix}^T,
\end{aligned}
$$

and

$$
\begin{aligned}
A_2 &= \begin{bmatrix} -0.2 & 0 \\ 0.1 & 0.1 \end{bmatrix}, \\
A_{d2} &= \begin{bmatrix} 0.05 & -0.1 \\ 0 & 0.15 \end{bmatrix}, \\
B_2 &= \begin{bmatrix} 0.1 & -1 \end{bmatrix}^T, \\
C_2 &= \begin{bmatrix} -0.19 & 0.17 \end{bmatrix}, \\
C_{d2} &= \begin{bmatrix} 0 & 0.017 \end{bmatrix}, \\
D_2 &= 0.015, \\
G_2 &= \begin{bmatrix} 0.2 & 0.1 \end{bmatrix}^T.
\end{aligned}
$$

Our objective is to design a filter such that the filtering error system under arbitrary switching signal is asymptotically stable with an H_∞ prescribed attenuation level γ. Let $\gamma = 1.2$, by using Matlab LMI toolbox to solve condition (2.14), a feasible solution is obtained. Then, by virtue of Theorem 2.7, the parameters of the filters are as follows:

$$A_{f1} = \begin{bmatrix} 0.0698 & -0.1113 \\ -0.1416 & 0.2312 \end{bmatrix},$$
$$B_{f1} = \begin{bmatrix} -1.4127 \\ -1.0539 \end{bmatrix},$$
$$C_{f1} = \begin{bmatrix} -0.082 & -0.2227 \end{bmatrix},$$

and

$$A_{f2} = \begin{bmatrix} -0.0715 & -0.0425 \\ 0.0496 & 0.0475 \end{bmatrix},$$
$$B_{f2} = \begin{bmatrix} -0.4480 \\ -0.2254 \end{bmatrix},$$
$$C_{f2} = \begin{bmatrix} -0.0881 & -0.0878 \end{bmatrix}.$$

In order to provide the figures, the initial condition is selected as $x_0 = \begin{bmatrix} 1 & 2 \end{bmatrix}^T$ and $\hat{x}_0 = \begin{bmatrix} 0 & 0 \end{bmatrix}^T$ and the disturbance is $w_k = 0.9 * exp(-0.5k) * cos(0.1k)$. Fig. 2.1 shows arbitrary switching signal. Fig. 2.2 demonstrates the estimated output

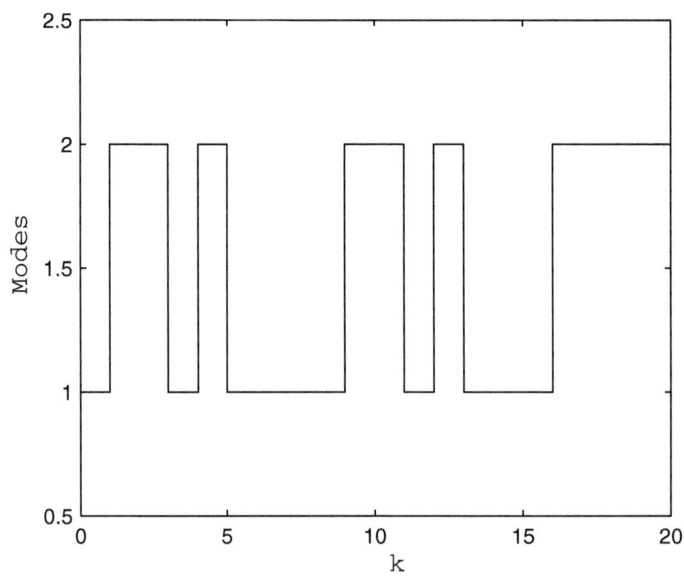

Fig. 2.1 Switching signal

signal z_k and the output of the filter $\hat{z}_k$. From the simulation results, it is seen that the designed filter satisfies the requirements.

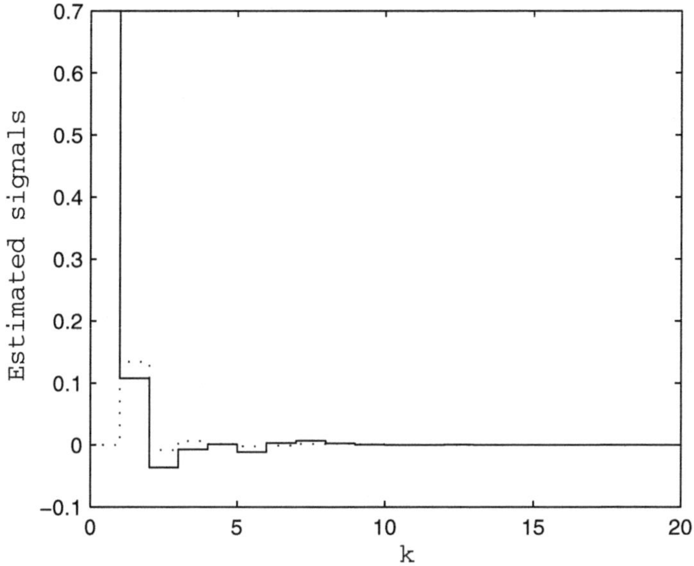

Fig. 2.2 Signal to be estimated z_k (solid) and filter output $\hat{z}_k$ (dashed)

2.6 Summary

We have studied the problem of H_∞ filtering for discrete-time switched linear system with a constant delay. Switched linear filters are such that the filtering error systems under arbitrary switching signals are stable with an H_∞ noise attenuation level. Switched Lyapunov function approach has been utilized to analyze the stability of the filtering error systems. A new linearization technique has been proposed to construct the parameters of filters. Simulation results have been provided to verify the effectiveness of the developed method.

3

Delay-Dependent H_∞ Filtering for Discrete-Time Switched Systems with a Time-Varying Delay

This chapter deals with the problem of H_∞ filtering for discrete-time switched delay systems. The time delay under consideration varies in a time interval, whose lower bound is allowed to be a positive integer. By utilizing Switched Lyapunov function approach and free-weighting matrix method, delay-dependent conditions for the existence of switched linear filters are derived and the parameter design is given in terms of LMIs.

3.1 Introduction

In the previous chapter, delay-independent conditions for the H_∞ filter design for discrete-time switched systems with state delays are obtained. The information on the size of delays is not utilized such that the results are applicable to the systems with arbitrary size of a constant time delay. But delay-dependent conditions are generally less conservative than the delay-independent ones, especially when the size of the delays is small. In general, a constant time delay is conservative since the real environment factors result in the variation of a time delay. It is well known that there are systems which are stable with some nonzero delay, but are unstable without delay [37]. For such case, if there is a time-varying perturbation on the nonzero delay, it is of great significance to consider the system with an interval time-varying delay. In [72], delay-dependent sufficient conditions for the solvability of the H_∞ filtering problem for discrete-time switched systems with state delays are presented. However, the conditions in [72] obtained are LMIs with a non-convex constraint, which are difficult to find the numerical solutions and lead to the heavy computation [26]. Consequently, strict LMI-based formulation is actually desired to easily find the numerical solution and avoid the heavy computation induced by the non-convex constraints. This motivates us to study this interesting and challenging problem.

In this chapter, we will improve the results established in [72] to obtain strict LMI-based conditions for the existence of the desired filters. Some linearization techniques are employed and additional slack variables are introduced to construct

D. Wang et al.: *Robust Filtering & Fault Detect. of Switched Delay Systems*, LNCIS 445, pp. 35–46.
DOI: 10.1007/978-3-642-37685-6_3

an admissible filter. Finally, a simulation example is provided to demonstrate the effectiveness of the proposed method.

The rest of this chapter is organized as follows. In Section 3.2, the H_∞ filtering performance is analyzed. Sufficient conditions on the existence of the H_∞ filters for discrete-time switched systems with state delays are derived via strict LMI formulation, and the desired filters are constructed in Section 3.3. To demonstrate the validity of the proposed approach, an example is given in Section 3.4 which is followed by a conclusion in Section 3.5.

3.2 Problem Formulation

Consider the following discrete-time switched systems with a state delay:

$$\begin{aligned} x_{k+1} &= A_\sigma x_k + A_{d\sigma} x_{k-d_k} + B_\sigma w_k, \\ y_k &= C_\sigma x_k + C_{d\sigma} x_{k-d_k} + D_\sigma w_k, \\ z_k &= G_\sigma x_k + G_{d\sigma} x_{k-d_k} + E_\sigma w_k, \\ x_l &= \phi_l, l = -d_M, -d_M+1, \ldots, 0. \end{aligned} \tag{3.1}$$

where $x_k \in R^n$ is the state, $y_k \in R^r$ is the measured output, $z_k \in R^q$ is the signal to be estimated. $w_k \in R^p$ is the disturbance input, which is assumed to belong to $l_2[0,\infty)$. $\sigma : [0, \infty) \to \mathscr{N}$ is the switching signal which specifies which subsystem is activated at the switching instant. $\mathscr{N}$ and d_k denote an index set of subsystems and the state delay, respectively. d_k is assumed to be an interval time-varying delay in the state satisfying $0 < d_m \le d_k \le d_M$, where d_m and d_M are scalars representing the lower and the upper bounds of the time delay, respectively. The matrices A_p, A_{dp}, B_p, C_p, C_{dp}, D_p, G_p, G_{dp} and $E_p, p \in \mathscr{N}$ of the pth subsystem are of the appropriate dimensions.

We are interested in designing a full-order switched filter described by

$$\begin{aligned} \hat{x}_{k+1} &= A_{f\sigma} \hat{x}_k + B_{f\sigma} y_k, \\ \hat{z}_k &= C_{f\sigma} \hat{x}_k + D_{f\sigma} y_k, \end{aligned} \tag{3.2}$$

where $\hat{x}_k \in R^n$ and $\hat{z}_k \in R^q$. σ is defined in (3.1). The matrices A_{fp}, B_{fp}, C_{fp} and D_{fp} are the parameters of the pth subsystem of filter (3.2) to be determined.

Remark 3.1. As assumed in the previous chapter, the switching signals are assumed to be unknown a priori, but their instantaneous values are available in real-time implementation. Moreover, the same switching signals are used in system (3.1) and filter (3.2). Thus, when the switching signal varies, the switchings between the subsystems in system (3.1) and filter (3.2) happen simultaneously.

Here, in view of (3.1) and (3.2), the filtering error systems are described by

$$\begin{aligned} \tilde{x}_{k+1} &= \tilde{A}_\sigma \tilde{x}_k + \tilde{A}_{d\sigma} E \tilde{x}_{k-d_k} + \tilde{B}_\sigma w_k, \\ \tilde{z}_k &= \tilde{C}_\sigma \tilde{x}_k + \tilde{C}_{d\sigma} E \tilde{x}_{k-d_k} + \tilde{D}_\sigma w_k, \end{aligned} \tag{3.3}$$

where

$$
\begin{aligned}
\tilde{x}_k^T &= \begin{bmatrix} x_k^T \ \hat{x}_k^T \end{bmatrix}, \\
\tilde{z}_k &= z_k - \hat{z}_k,\ E = \begin{bmatrix} I & 0 \end{bmatrix}, \\
\tilde{A}_\sigma &= \begin{bmatrix} A_\sigma & 0 \\ B_{f\sigma}C_\sigma & A_{f\sigma} \end{bmatrix}, \\
\tilde{A}_{d\sigma} &= \begin{bmatrix} A_{d\sigma} \\ B_{f\sigma}C_{d\sigma} \end{bmatrix}, \\
\tilde{B}_\sigma &= \begin{bmatrix} B_\sigma \\ B_{f\sigma}D_\sigma \end{bmatrix}, \\
C_\sigma &= \begin{bmatrix} G_\sigma - D_{f\sigma}C_\sigma & -C_{f\sigma} \end{bmatrix}, \\
\tilde{C}_{d\sigma} &= G_{d\sigma} - D_{f\sigma}C_{d\sigma}, \\
\tilde{D}_\sigma &= E_\sigma - D_{f\sigma}D_\sigma.
\end{aligned}
$$

The problem addressed in this paper is to design an H_∞ filter in the form of (3.2) such that the following specifications are met for filtering error system (3.3) under arbitrary switching signals:

(1) Filtering error system (3.3) is asymptotically stable when $w_k = 0$;
(2) Filtering error system (3.3) guarantees, under zero-initial condition, $\|\tilde{z}_k\|_2 \leq \gamma\|w_k\|_2$ for all nonzero $w_k \in l_2[0,\infty)$ and a given positive constant γ.

3.3 Stability and H_∞ Performance Analysis

In this section, the following Lemma provides an H_∞ performance analysis for filtering error system (3.3), which will play an important role in our filter design.

Lemma 3.2. *For a given scalar $\gamma > 0$, the problem of H_∞ filter design is solvable for system (3.1) under arbitrary switching signals, if there exist symmetric and positive definite matrices P_p, Q_l and Z_r, and matrices M_{pr}, S_{pr} and N_{pr}, $p \in \mathscr{N}, l \in \{1,2,3\}, r \in \{1,2\}$ such that the following inequality holds,*

$$
\begin{bmatrix} \Pi_1 & \Pi_2 \\ * & -\Pi_3 \end{bmatrix} < 0, \tag{3.4}
$$

where

$$
\Pi_1 = \begin{bmatrix} \Phi_{11} & \Phi_{12} & 0 \\ * & \Phi_{22} & 0 \\ * & * & -\gamma^2 I \end{bmatrix},
$$

$$
\Pi_2 = \begin{bmatrix} N_{p1} & -S_{p1} & \sqrt{d_M}M_{p1} & \sqrt{\bar{d}-1}S_{p1} & \sqrt{\bar{d}}N_{p1} & \tilde{C}_p^T & \tilde{A}_p^T & E^T(A_p - I)^T \\ N_{p2} & -S_{p2} & \sqrt{d_M}M_{p2} & \sqrt{\bar{d}-1}S_{p2} & \sqrt{\bar{d}}N_{p2} & \tilde{C}_{dp}^T & \tilde{A}_{dp}^T & A_{dp}^T \\ 0 & 0 & 0 & 0 & 0 & \tilde{D}_p^T & \tilde{B}_p^T & B_p^T \end{bmatrix},
$$

$$
\begin{aligned}
&\Pi_3 = diag\left\{ Q_2 \; Q_3 \; Z_1 \; Z_3 \; Z_2 \; I \; P_q^{-1} \; Z_4^{-1} \right\},\\
&\Phi_{11} = -P_p + \bar{d}E^T Q_1 E + E^T(Q_2+Q_3)E + E^T M_{p1}^T + M_{p1}E,\\
&\Phi_{12} = E^T M_{p2}^T - M_{p1} + S_{p1} - N_{p1},\\
&\Phi_{22} = -Q_1 + S_{p2} + S_{p2}^T - M_{p2} - M_{p2}^T - N_{p2} - N_{p2}^T,\\
&Z_3 = Z_1 + Z_2, Z_4 = d_M Z_1 + \bar{d}Z_2, \bar{d} = d_M - d_m + 1.
\end{aligned}
$$

Proof. The following Lyapunov functional is constructed by:

$$
V_{\sigma k} = V_{\sigma 1k} + V_{2k} + V_{3k} + V_{4k} + V_{5k} + V_{6k}, \tag{3.5}
$$

where

$$
\begin{aligned}
V_{\sigma 1k} &= \tilde{x}_k^T P_\sigma \tilde{x}_k,\\
V_{2k} &= \sum_{l=k-d_k}^{k-1} x_l^T Q_1 x_l + \sum_{j=-d_M+1}^{-d_m} \sum_{l=k+j}^{k-1} x_l^T Q_1 x_l,\\
V_{3k} &= \sum_{l=k-d_m}^{k-1} x_l^T Q_2 x_l,\\
V_{4k} &= \sum_{l=k-d_M}^{k-1} x_l^T Q_3 x_l,\\
V_{5k} &= \sum_{j=-d_M}^{-1} \sum_{l=k+j}^{k-1} \eta_l^T Z_1 \eta_l,\\
V_{6k} &= \sum_{j=-d_M}^{-d_m} \sum_{l=k+j}^{k-1} \eta_l^T Z_2 \eta_l,\\
\eta_l &= x_{l+1} - x_l.
\end{aligned}
$$

Taking the forward difference $\Delta V_{\sigma k} = V_{\sigma k+1} - V_{\sigma k}$ along the trajectory of system (3.3), we get

$$
\Delta V_{\sigma 1k} = \tilde{x}_{k+1}^T P_p \tilde{x}_{k+1} - \tilde{x}_k^T P_q \tilde{x}_k, \tag{3.6}
$$

and

$$
\begin{aligned}
\Delta V_{2k} = {} & \bar{d} x_k^T Q_1 x_k - x_{k-d_k}^T Q_1 x_{k-d_k},\\
& + \sum_{l=k+1-d_{k+1}}^{k-1} x_l^T Q_1 x_l - \sum_{l=k+1-d_k}^{k-1} x_l^T Q_1 x_l - \sum_{l=k+1-d_M}^{k-d_m} x_l^T Q_1 x_l.
\end{aligned}
$$

Noting that $d_m \le d_k \le d_M$, we have

$$
\sum_{l=k+1-d_{k+1}}^{k-1} x_l^T Q_1 x_l \le \sum_{l=k+1-d_k}^{k-1} x_l^T Q_1 x_l + \sum_{l=k+1-d_M}^{k-d_m} x_l^T Q_1 x_l.
$$

Then,

$$\Delta V_{2k} \leq \bar{d}x_k^T Q_1 x_k - x_{k-d_k}^T Q_1 x_{k-d_k}. \tag{3.7}$$

Similarly, we have

$$\Delta V_{3k} = x_k^T Q_2 x_k - x_{k-d_m}^T Q_2 x_{k-d_m}, \tag{3.8}$$

$$\Delta V_{4k} = x_k^T Q_3 x_k - x_{k-d_M}^T Q_3 x_{k-d_M}, \tag{3.9}$$

$$\Delta V_{5k} = d_M \eta_k^T Z_1 \eta_k - \sum_{l=k-d_k}^{k-1} \eta_l^T Z_1 \eta_l - \sum_{l=k-d_M}^{k-d_k-1} \eta_l^T Z_1 \eta_l, \tag{3.10}$$

$$\Delta V_{6k} = \bar{d}\eta_k^T Z_2 \eta_k - \sum_{l=k-d_k}^{k-d_m} \eta_l^T Z_2 \eta_l - \sum_{l=k-d_M}^{k-d_k-1} \eta_l^T Z_2 \eta_l. \tag{3.11}$$

By means of the free-weighted matrices method [123, 41], the following equations are introduced:

$$2\xi_k^T M_p \left[x_k - x_{k-d_k} - \sum_{l=k-d_k}^{k-1} \eta_l \right] = 0, \tag{3.12}$$

$$2\xi_k^T S_p \left[x_{k-d_k} - x_{k-d_M} - \sum_{l=k-d_M}^{k-d_k-1} \eta_l \right] = 0, \tag{3.13}$$

$$2\xi_k^T N_p \left[x_{k-d_m} - x_{k-d_k} - \sum_{l=k-d_k}^{k-d_m} \eta_l \right] = 0, \tag{3.14}$$

where

$$\begin{aligned}
\xi_k &= \left[\tilde{x}_k^T \; x_{k-d_k}^T \; x_{k-d_m}^T \; x_{k-d_M}^T \right]^T, \\
M_p &= \left[M_{p1}^T \; M_{p2}^T \; 0 \; 0 \right]^T, \\
S_p &= \left[S_{p1}^T \; S_{p2}^T \; 0 \; 0 \right]^T, \\
N_p &= \left[N_{p1}^T \; N_{p2}^T \; 0 \; 0 \right]^T.
\end{aligned}$$

From (3.10) and (3.12), we have

$$
\begin{aligned}
&\sum_{l=k-d_k}^{k-1} \eta_l^T Z_1 \eta_l + 2\xi_k^T M_p \sum_{l=k-d_k}^{k-1} \eta_l \\
&+ \sum_{l=k-d_k}^{k-1} \xi_k^T M_p Z_1^{-1} M_p^T \xi_k - \sum_{l=k-d_k}^{k-1} \xi_k^T M_p Z_1^{-1} M_p^T \xi_k \\
&= \sum_{l=k-d_k}^{k-1} \left[\xi_k^T M_p + \eta_l^T Z_1\right] Z_1^{-1} \left[M_p^T \xi_k + Z_1 \eta_l\right] - d_k \xi_k^T M_p Z_1^{-1} M_p^T \xi_k \\
&\geq -d_M \xi_k^T M_p Z_1^{-1} M_p^T \xi_k. \qquad (3.15)
\end{aligned}
$$

Similarly, we can get

$$
\begin{aligned}
&\sum_{l=k-d_M}^{k-d_k-1} \eta_l^T Z_1 \eta_l + \sum_{l=k-d_M}^{k-d_k-1} \eta_l^T Z_2 \eta_l + 2\xi_k^T S_p \sum_{l=k-d_M}^{k-d_k-1} \eta_l \\
&+ \sum_{l=k-d_M}^{k-d_k-1} \xi_k^T S_p (Z_1 + Z_2)^{-1} S_p^T \xi_k \\
&- \sum_{l=k-d_M}^{k-d_k-1} \xi_k^T S_p (Z_1 + Z_2)^{-1} S_p^T \xi_k \\
&\geq -(\bar{d} - 1) \xi_k^T S_p (Z_1 + Z_2)^{-1} S_p^T \xi_k, \qquad (3.16)
\end{aligned}
$$

and

$$
\begin{aligned}
&\sum_{l=k-d_k}^{k-d_m} \eta_l^T Z_2 \eta_l + 2\xi_k^T N_p \sum_{l=k-d_k}^{k-d_m} \eta_l \\
&+ \sum_{l=k-d_k}^{k-d_m} \xi_k^T N_p Z_2^{-1} N_p^T \xi_k - \sum_{l=k-d_k}^{k-d_m} \xi_k^T N_p Z_2^{-1} N_p^T \xi_k \\
&\geq -\bar{d} \xi_k^T N_p Z_2^{-1} N_p^T \xi_k. \qquad (3.17)
\end{aligned}
$$

First of all, for system (3.3) with $w_k = 0$, by (3.6)-(3.11), (3.15)-(3.17) and Schur complement, it is obtained from (3.4) that

$$
\begin{aligned}
&\Delta V_{\sigma k} + 2\xi_k^T M_p \left[x_k - x_{k-d_k} - \sum_{l=k-d_k}^{k-1} \eta_l\right] \\
&+2\xi_k^T S_p \left[x_{k-d_k} - x_{k-d_M} - \sum_{l=k-d_M}^{k-d_k-1} \eta_l\right] \\
&+ 2\xi_k^T N_p \left[x_{k-d_m} - x_{k-d_k} - \sum_{l=k-d_k}^{k-d_m-1} \eta_l\right] \\
&< -\rho \left\|\xi_k\right\|^2,
\end{aligned}
$$

for a sufficiently small $\rho > 0$ and $\xi_k \neq 0$, which establishes the asymptotical stability of system (3.3) with $w_k = 0$.

Secondly, for any nonzero $w_k \in l_2[0,\infty)$, we consider the following performance index:

$$J = \sum_{k=0}^{K-1} \left[\tilde{z}_k^T \tilde{z}_k - \gamma^2 w_k^T w_k\right],$$

where K is an arbitrary positive integer.

Under zero-initial condition $\tilde{x}_0 = 0$, one has

$$\begin{aligned} J &= \sum_{k=0}^{K-1} \left[\tilde{z}_k^T \tilde{z}_k - \gamma^2 w_k^T w_k + \Delta V_{\sigma k}\right] - V_{\sigma K} \\ &\leq \sum_{k=0}^{K-1} \left[\tilde{z}_k^T \tilde{z}_k - \gamma^2 w_k^T w_k + \Delta V_{\sigma k}\right]. \end{aligned}$$

Utilizing (3.6)-(3.11) and (3.15)-(3.17), we have

$$\begin{aligned} &\tilde{z}_k^T \tilde{z}_k - \gamma^2 w_k^T w_k + \Delta V_{\sigma k} \\ &= \tilde{\xi}_k^T (\Upsilon_p + d_M M_p Z_1^{-1} M_p^T + (\bar{d}-1) S_p (Z_1+Z_2)^{-1} S_p^T + \bar{d} N_p Z_2^{-1} N_p^T) \tilde{\xi}_k \\ &+ \tilde{z}_k^T \tilde{z}_k + x_{k+1}^T P_q x_{k+1} + \eta_k^T (d_M Z_1 + \bar{d} Z_2) \eta_k, \end{aligned} \tag{3.18}$$

where

$$\tilde{\xi}_k = \left[\tilde{x}_k^T \; x_{k-d_k}^T \; w_k^T \; x_{k-d_m}^T \; x_{k-d_M}^T\right]^T,$$

$$\Upsilon_p = \begin{bmatrix} \Phi_{11} & \Phi_{12} & 0 & N_{p1} & -S_{p1} \\ * & \Phi_{22} & 0 & N_{p2} & -S_{p2} \\ * & * & -\gamma^2 I & 0 & 0 \\ * & * & * & -Q_2 & 0 \\ * & * & * & * & -Q_3 \end{bmatrix}.$$

It follows from (3.4) and Schur complement that $\tilde{z}_k^T \tilde{z}_k - \gamma^2 w_k^T w_k + \Delta V_{\sigma k} < 0$, which implies $J < 0$. Then, one has that for any nonzero $w_k \in l_2[0,\infty)$, $\|\tilde{z}_k\|_2 < \gamma \|w_k\|_2$, which completes the proof.

Remark 3.3. As switched Lyapunov functional method [10] is used in (3.6), $p = q$ shows the occurrence of no switching while $p \neq q$ represents the switching from the pth subsystem to the qth subsystem. Therefore, the switching signals are arbitrary.

Remark 3.4. Notice that P_p is dealt with by switched Lyapunov functional method and others are common Lyapunov matrices. The main reason is that the existence of time delay leads to difficultly deal with the time difference terms of Lyapunov function associated with time delay when such Lyapunov function is selected to analyze and synthesize switched delay systems. This is also pointed out in [144]. However, the utilization of common Lyapunov matrices overcomes the difficulty and makes it feasible to obtain strict LMI formulation for the solvability of H_∞ filtering problem.

3.4 H_∞ Filter Design

In this section, a sufficient condition for the existence of a stable filter assuring an H_∞ performance index for system (3.1) will be given.

Theorem 3.5. *Let $\gamma > 0$ be a given scalar, an desired H_∞ filter can be designed for system (3.1), if there exist symmetric and positive definite matrices P_p, Q_l and Z_r, and matrices $\tilde{\Omega}$, X_r, F_{sp}, M_{pr}, S_{pr} and N_{pr}, $p \in \mathscr{N}, l \in \{1,2,3\}, s \in \{1,2,3,4\}, r \in \{1,2\}$ such that the following inequality holds,*

$$\begin{bmatrix} \Pi_1 & \Pi_4 \\ * & -\Pi_5 \end{bmatrix} < 0, \tag{3.19}$$

where

$$\Pi_4 = \begin{bmatrix} N_{p1} & -S_{p1} & \sqrt{d_M}M_{p1} & \sqrt{\bar{d}-1}S_{p1} & \sqrt{\bar{d}}N_{p1} & \Phi_{13}^T & \Phi_{14}^T & E^T(A_p-I)^T\tilde{\Omega}^T \\ N_{p2} & -S_{p2} & \sqrt{d_M}M_{p2} & \sqrt{\bar{d}-1}S_{p2} & \sqrt{\bar{d}}N_{p2} & \Phi_{23}^T & \Phi_{24}^T & A_{dp}^T\tilde{\Omega}^T \\ 0 & 0 & 0 & 0 & 0 & \Phi_{33}^T & \Phi_{34}^T & B_p^T\tilde{\Omega}^T \end{bmatrix},$$

$$\Pi_5 = diag\left\{ Q_2 \; Q_3 \; Z_1 \; Z_3 \; Z_2 \; I \; -\hat{P}_q \; -\bar{Z}_4 \right\},$$

$$\hat{P}_q = P_q - \left(\Phi_{44}^T + \Phi_{44}\right),$$

$$\bar{Z}_4 = Z_4 - \left(\tilde{\Omega}^T + \tilde{\Omega}\right),$$

$$\Phi_{13} = \left[G_p - F_{4p}C_p \;\; -F_{3p} \right],$$

$$\Phi_{23} = G_{dp} - F_{4p}C_{dp},$$

$$\Phi_{33} = E_p - F_{4p}D_p,$$

$$\Phi_{14} = \begin{bmatrix} A_p^T X_1^T - C_p^T F_{2p}^T & -A_p^T X_2^T + C_p^T F_{2p}^T \\ -F_{1p}^T & F_{1p}^T \end{bmatrix},$$

$$\Phi_{24} = \left[A_{dp}^T X_1^T - C_{dp}^T F_{2p}^T \;\; -A_{dp}^T X_2^T + C_{dp}^T F_{2p}^T \right],$$

$$\Phi_{34} = \left[B_p^T X_1^T - D_p^T F_{2p}^T \;\; -B_p^T X_2^T + D_p^T F_{2p}^T \right],$$

$$\Phi_{44} = \begin{bmatrix} X_1 & -X_2 \\ * & X_2 \end{bmatrix}.$$

Then, the parameters of the pth filter in the form of (3.2) can be given by

$$\begin{bmatrix} A_{fp} & B_{fp} \\ C_{fp} & D_{fp} \end{bmatrix} = \begin{bmatrix} X_2^{-1}F_{1p} & X_2^{-1}F_{2p} \\ F_{3p} & F_{4p} \end{bmatrix}. \tag{3.20}$$

Proof. By Lemma 3.2, system (3.3) under arbitrary switching signals is asymptotically stable when $w_k = 0$ and, under zero-initial conditions, guarantees $\|\tilde{z}_k\|_2 < \gamma\|w_k\|_2$ for all nonzero $w_k \in l_2[0,\infty)$, if LMI (3.4) holds.

Note that from (3.19),

$$P_q - \left(\Phi_{44}^T + \Phi_{44}\right) < 0,$$

which means that X_1 and X_2 are nonsingular.

Now, introduce

$$F_{1p} \triangleq X_2 A_{fp},\ F_{2p} \triangleq X_2 B_{fp},$$
$$F_{3p} \triangleq C_{fp},\ F_{4p} \triangleq D_{fp}. \tag{3.21}$$

By using (3.3), (3.19) and (3.21), we can get

$$\Phi_{14} = \Phi_{44}\tilde{A}_p,\ \Phi_{24} = \Phi_{44}\tilde{A}_{dp},\ \Phi_{34} = \Phi_{44}\tilde{B}_p. \tag{3.22}$$

Substituting (3.22) into (3.19) yields

$$\begin{bmatrix} \Pi_1 & \Pi_6 \\ * & -\Pi_7 \end{bmatrix} < 0, \tag{3.23}$$

where

$$\Pi_6 = \begin{bmatrix} N_{p1} & -S_{p1} & \sqrt{d_M}M_{p1} & \sqrt{\bar{d}-1}S_{p1} & \sqrt{\bar{d}}N_{p1} & \tilde{C}_p^T & \tilde{A}_p^T\Phi_{44}^T & E^T(A_p-I)^T\tilde{\Omega}^T \\ N_{p2} & -S_{p2} & \sqrt{d_M}M_{p2} & \sqrt{\bar{d}-1}S_{p2} & \sqrt{\bar{d}}N_{p2} & \tilde{C}_{dp}^T & \tilde{A}_{dp}^T\Phi_{44}^T & A_{dp}^T\tilde{\Omega}^T \\ 0 & 0 & 0 & 0 & 0 & \tilde{D}_p^T & \tilde{B}_p^T\Phi_{44}^T & B_p^T\tilde{\Omega}^T \end{bmatrix},$$

$$\Pi_7 = diag\left\{ Q_2,\ Q_3,\ Z_1,\ Z_3,\ Z_2 \nu I,\ -\bar{P}_q,\ -\bar{Z}_4 \right\},$$
$$\bar{P}_q = P_q - \left(\Phi_{44}^T + \Phi_{44}\right).$$

On the other hand, from

$$(P_q - \Phi_{44})P_q^{-1}(P_q - \Phi_{44})^T \geq 0,$$

we know that

$$-\Phi_{44}P_q^{-1}\Phi_{44}^T \leq P_q - \left(\Phi_{44} + \Phi_{44}^T\right). \tag{3.24}$$

Similarly, one has

$$-\tilde{\Omega}Z_4^{-1}\tilde{\Omega}^T \leq Z_4 - \left(\tilde{\Omega} + \tilde{\Omega}^T\right). \tag{3.25}$$

Hence, from (3.19), by means of (3.24) and (3.25), we can get

$$\begin{bmatrix} \Pi_1 & \Pi_6 \\ * & -\Pi_8 \end{bmatrix} < 0, \tag{3.26}$$

with $\Pi_8 = diag\left\{ Q_2,\ Q_3,\ Z_1,\ Z_3,\ Z_2,\ I,\ \Phi_{44}P_q^{-1}\Phi_{44}^T,\ \tilde{\Omega}Z_4^{-1}\tilde{\Omega}^T \right\}$.

Performing a congruence transformation to (3.26) via

$$diag\left\{I,\ I,\ I,\ I,\ I,\ I,\ I,\ I,\ I,\ \Phi_{44}^{-T},\ \tilde{\Omega}^{-T}\right\},$$

yields (3.4). Therefore, (3.19) holds, which implies that (3.4) holds. Meanwhile, we know from (3.22) that the parameters of an admissible filter are given by (3.20). The proof is completed.

Remark 3.6. In [15], delay-independent conditions are obtained for discrete-time switched systems with constant delays. In [72], a delay-dependent condition is given for discrete-time switched systems with time-varying delays. Clearly, the criteria in [72] are less conservative than those in [15] when time delays are small [40, 41]. Notice from Remark 3.2 in [72] that $\mathscr{S}_i = \mathscr{W}_i^{-1}$ is used to decouple the Lyapunov matrices $\mathscr{W}_i^{-1}$ from the system matrices $\tilde{A}_i$, $\tilde{A}_{di}$, and thus derive Theorem 3.5. However, the simultaneous existence of $\mathscr{S}_i$ and $\mathscr{W}_i$ in (3.16) causes that the filter design conditions in [72] is LMIs with a non-convex constraint $\mathscr{S}_i\mathscr{W}_i = 1$. This non-convex constraint makes it difficult to find the numerical solution, which is NP-hard [26]. Thus, it is desired to obtain strict LMI conditions which are easy to be solved numerically and avoid the heavy calculation. In Theorem 3.5, by introducing Φ_{44} and using (3.24), some product terms between the Lyapunov matrices $\bar{P}_q^{-1}$ and the system matrices $\tilde{A}_p$, $\tilde{A}_{dp}$ and $\tilde{B}_p$ are removed. Similarly, $\tilde{\Omega}$ and (3.25) are employed to eliminate the cross coupling of the Lyapunov matrices $\bar{Z}_4^{-1}$ and the system matrices A_p, A_{dp} and B_p. These imply that it is tractable to obtain strict LMI-based conditions to design a desired filter in Theorem 3.5.

3.5 An Illustrative Example

In this section, an example is provided to illustrate the effectiveness of the proposed method.

Consider discrete-time switched system (3.1) consisting of two subsystems with parameters:

$$\begin{aligned}
A_1 &= \begin{bmatrix} 0.15 & 0 \\ 0.1 & 0.2 \end{bmatrix}, \\
A_{d1} &= \begin{bmatrix} 0.1 & 0.3 \\ 0.2 & 0.15 \end{bmatrix}, \\
B_1 &= \begin{bmatrix} 0 & 0.1 \end{bmatrix}^T, \; C_1 = \begin{bmatrix} 0 & 0.2 \end{bmatrix}, \\
C_{d1} &= \begin{bmatrix} 0 & 0.3 \end{bmatrix}, \; G_1 = \begin{bmatrix} 0.3 & 0 \end{bmatrix}, \\
G_{d1} &= \begin{bmatrix} 0.2 & 0 \end{bmatrix}, \\
D_1 &= 0.2, \; E_1 = 0.3.
\end{aligned}$$

and

$$\begin{aligned}
A_2 &= \begin{bmatrix} 0.15 & 0.1 \\ 0.3 & 0.25 \end{bmatrix}, \\
A_{d2} &= \begin{bmatrix} 0.15 & 0.1 \\ 0.3 & 0.25 \end{bmatrix}, \\
B_2 &= \begin{bmatrix} 0 & 0.15 \end{bmatrix}^T, \; C_2 = \begin{bmatrix} 0 & 0.15 \end{bmatrix}, \\
C_{d2} &= \begin{bmatrix} 0 & 0.25 \end{bmatrix}, \; G_2 = \begin{bmatrix} 0.15 & 0 \end{bmatrix}, \\
G_{d2} &= \begin{bmatrix} 0.2 & 0 \end{bmatrix}, \\
D_2 &= 0.2, \; E_2 = 0.15.
\end{aligned}$$

In addition, the lower and upper bounds of the interval time-varying delay are assumed to be $d_m = 1$, $d_M = 3$. For a given $\gamma = 0.8$, by solving (3.19), we can have a feasible solution. For simplicity, here, we only present the parameter matrices of the filter by (3.20) as follows:

$$\left[\begin{array}{c|c} A_{f1} & B_{f1} \\ \hline C_{f1} & D_{f1} \end{array}\right] = \left[\begin{array}{cc|c} 0.0656 & -0.0532 & 0.7398 \\ 0.0812 & 0.0244 & 1.0333 \\ \hline -0.0868 & 0.1549 & 1.0042 \end{array}\right],$$

$$\left[\begin{array}{c|c} A_{f2} & B_{f2} \\ \hline C_{f2} & D_{f2} \end{array}\right] = \left[\begin{array}{cc|c} 0.1052 & 0.0680 & 0.2783 \\ 0.1736 & 0.1167 & 1.1251 \\ \hline -0.0699 & 0.0569 & 0.5927 \end{array}\right].$$

We assume that time in samples is 100 and the energy bounded disturbance input is $w_k = 0.1e^{-0.4k}cos(0.2\pi k)$. The switching signal can be generated randomly, which is shown in Fig. 3.1. Note that the switching laws are arbitrary, which demonstrates that subsystem 1 or 2 is activated arbitrarily. Furthermore, given the zero-initial conditions and $1 \leq d_k \leq 3$, the simulation results of the system and the filter are given in Fig. 3.2, which shows the simulation curves of the signal z_k and $\hat{z}_k$, respectively. It is clearly observed from the simulation results above that for given energy bounded disturbance, the filtering error system is asymptotically stable with an H_∞ norm bound, which thereby implies the effectiveness of our filter design method.

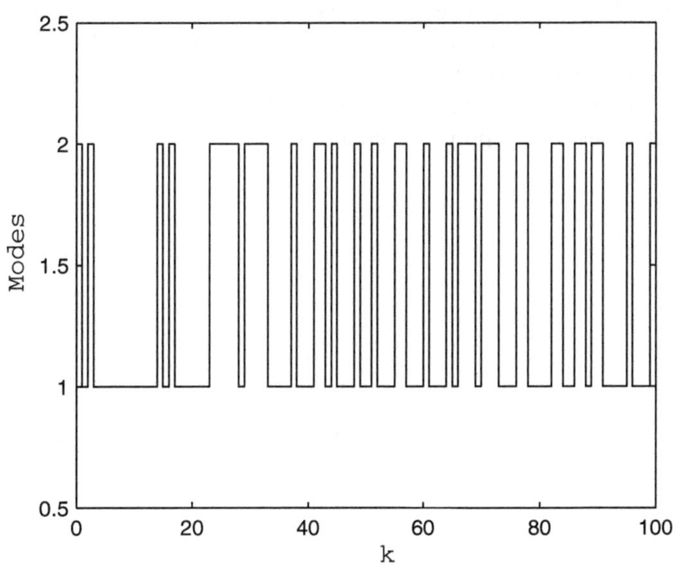

Fig. 3.1 Switching signal

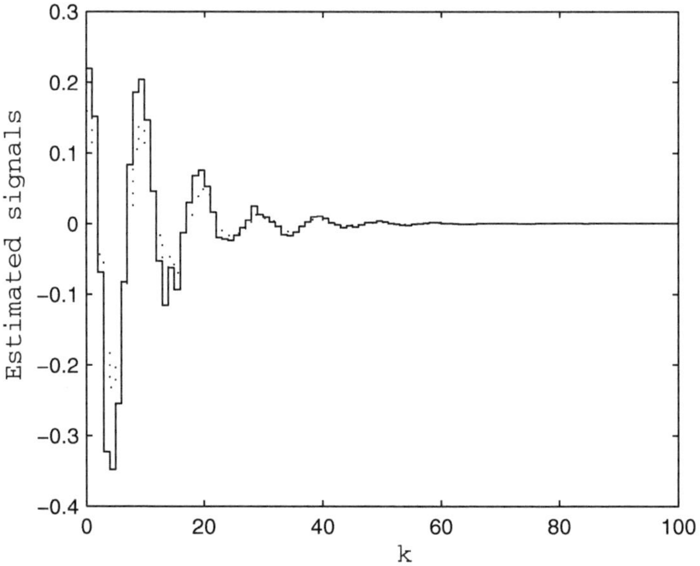

Fig. 3.2 Signal to be estimated z_k (solid) and filter output $\hat{z}_k$ (dashed)

3.6 Summary

The delay-dependent H_∞ filtering problem for discrete-time switched systems with a state delay has been investigated. By using switched Lyapunov functional approach in combining with Lyapunov-Krasovskii functional method, a sufficient condition for the solvability of this problem has been established in terms of strict LMIs and the admissible filter has been constructed. Neither model transformation nor bounding technique for cross terms has been employed. An example has been given to show the effectiveness of the proposed methods.

4

Delay-Dependent Exponential H_∞ Criteria for Discrete-Time Switched Delay Systems

This chapter is concerned with the problems of delay-dependent exponential H_∞ filtering and model reduction for discrete-time switched delay systems under average dwell time (ADT) switching signals. First, by introducing a proper factor to construct a novel Lyapunov-Krasovskii function and using ADT approach, sufficient conditions for the solvability of exponential H_∞ filtering problem, dependent on the upper and lower bounds of the time-varying delay, are obtained in terms of linear matrix inequalities (LMIs). The second objective is to construct a reduced-order model, which ensures that the resulting error system under switching signal with ADT is exponentially stable with an H_∞ norm bound. A weighting factor α is introduced to present sufficient conditions on the existence of reduced-order model in terms of strict LMIs, which lessen the computation complexity. Two numerical examples are presented to demonstrate the effectiveness of the developed results.

4.1 Delay-Dependent Exponential H_∞ Filtering

In previous two chapters, the switched Lyapunov function approach is employed to present sufficient conditions on the existence of H_∞ filter under arbitrary switching signals for switched systems with a constant delay and an interval time-varying delay, respectively. One of the shortcoming of arbitrary switching signals is that can lead to high-speed switching in a limited time interval and to generate the buffeting damaging the systems. For example, in a digital network, the finite transmission rate of the real-time data in the physical device limits the switching frequency. Therefore, the design of a switching law has to consider preventing the rapid switching or the chattering. The design of the low-frequency switching rate is difficult. Even for a simple linear time-invariant system, it is hard to present the constraints regarding the switching signals. An intuitive and simple idea is to restrict the minimum running time of each subsystem. Hence, a dwell time switching law is proposed as a slow switching law. The main idea is that, between two adjacent switchings, the minimum running time of the subsystems is not less than a given time interval. Further, based on this idea, average dwell time method is developed, namely, the average value of dwell

D. Wang et al.: *Robust Filtering & Fault Detect. of Switched Delay Systems*, LNCIS 445, pp. 47–74.
DOI: 10.1007/978-3-642-37685-6_4

times of all subsystems in two consecutive switching is not less than a given time constant. Dwell time and average dwell time methods have been employed widely. Average dwell time (ADT) methods as a special class of MLF methods are used to achieve the exponential stability for switched systems [42]. Also, an ADT method is employed to study the stability of switched systems with stable and unstable subsystems [138]. It is well known that time delays are the inherent features of many physical processes because of the limited speed of signals traveling among the components and the main sources of instability and poor performance. Therefore, switched systems with time delays have much wider engineering background and have been extensively studied recently. For some representative work on this general topic, to name a few, we refer readers to [120, 44, 58] and the references therein. In [58], uniform asymptotic stability of a class of linear switching systems with time delay is achieved. A switched digital control is constructed for continuous-time switched systems which is robust to the model uncertainties produced by unknown varying feedback delay [44]. An adaptive sliding mode controller for a continuous-time switched system with time-varying delay is designed by using the average dwell time approach and the piecewise Lyapunov function technique[120].

Up to now, H_∞ filtering in switched system framework is still in the early stage of development. H_∞ filtering problems for discrete-time switched polytopic linear systems free of time delay are addressed in [141, 146]. Delay-independent conditions on the existence of an H_∞ filter for discrete-time switched systems with time delays are obtained in [15]. Note that delay-dependent criteria are more appropriated than delay-independent criteria when time delay is small [89, 27, 41]. So, for discrete-time switched delay systems, delay-dependent conditions are derived to design H_∞ filters in [72]. All the results in [89, 141, 146, 72] are presented based on SLF approach. When SLF method is used to deal with switched systems with time-varying delay, the corresponding Lyapunov function will be involved with some common Lyapunov matrices for all subsystems[144]. ADT approach has been proven to be a powerful tool to design an exponential H_∞ filter, developed in [131], for discrete-time switched linear systems[141]. But time delays are not taken into account because it is difficult to deal with the time-delay terms when Lyapunov functions are used to analyze the stability of switched delay systems at the switching instant. Recently, an exponential H_∞ filter design approach is developed in [131] for uncertain Takagi-Sugeno fuzzy systems to guarantee the resultant filtering error system to be exponentially stable with a prescribed level of the l_2-induced gain from the noise signal to the estimation error. It is worth pointing out that the advantage of a filter with exponential stability in comparison with that with asymptotic stability lies in that the former provides fast convergence and desirable accuracy in terms of reasonable error covariance of the filtering process [92]. Meanwhile, it is also challenging to consider exponential H_∞ filtering problem for switched systems due to the complexity of mode switching. An exponential H_∞ filter for a class of polytopic uncertain discrete-time switched linear systems with average dwell time switching is designed by means of a Lyapunov function approach in [141]. To the best of our knowledge, little work has been conducted regarding exponential H_∞ filtering for switched state delay systems under ADT switching signals. This issue remains to be difficult and challenging due to the complexity of mode switching and the existence of state delay.

In this chapter, we construct a novel Lyapunov-Krasovskii function by introducing the factor α such that it is tractable to design an exponential H_∞ filter for switched delay systems under ADT switching signals. The advantages of introduction of α are as follows. First, it makes the time-varying delay terms tractable. Second, Lyapunov matrices concerning time delay in the constructed Lyapunov-Krasovskii function are not common but piecewise. Third, it makes the constructed Lyapunov-Krasovskii function more general in literature. Based on ADT method, we present delay-dependent sufficient conditions on the existence of exponential H_∞ filter for discrete-time switched systems with interval time-varying delays. A desired filter is designed by the linearization technique. A numerical example is given to illustrate the effectiveness of the proposed approach.

The rest of this chapter is organized as follows. In Section 4.1.1, the problem under consideration is formulated and in Section 4.1.2, the H_∞ filtering performance is analyzed. Sufficient conditions on the existence of the H_∞ filters for discrete-time switched systems with state delays are derived via an LMI formulation, and the desired filters are constructed in Section 4.1.3. To demonstrate the validity of the proposed approach, an example is given in Section 4.1.4 which is followed by a conclusion in Section 4.1.5.

4.1.1 Problem Formulation

Consider the following discrete-time switched systems with state delay:

$$\begin{aligned} x_{k+1} &= A_\sigma x_k + A_{d\sigma} x_{k-d_k} + B_\sigma w_k, \\ y_k &= C_\sigma x_k + C_{d\sigma} x_{k-d_k} + D_\sigma w_k, \\ z_k &= G_\sigma x_k + G_{d\sigma} x_{k-d_k} + E_\sigma w_k, \\ x_{\bar{s}} &= \phi_{\bar{s}},\ \bar{s} = -d_M,\ -d_M+1,\ \ldots,\ 0. \end{aligned} \tag{4.1}$$

where $x_k \in R^n$ is the state, $y_k \in R^r$ is the measured output, $w_k \in R^p$ is the disturbance input which belong to $l_2[0,\infty)$, $z_k \in R^q$ is the signal to be estimated. The piecewise constant function σ denoting σ_k for simplicity : $[0,\ \infty) \to \mathscr{P}=\{1,\cdots,N\}$ is a switching signal to specify, at each time instant k, $\sigma_k = i$ means that the ith subsystem is activated. The ith subsystem is denoted by constant matrices A_i, A_{di}, B_i, C_i, C_{di}, D_i, G_i, G_{di} and E_i of appropriate dimensions. d_k is assumed to be an interval time-varying delay in the state satisfying $d_m \le d_k \le d_M$, where d_m and d_M are two known constant positive integers representing the lower and the upper bounds of the time-varying delay, respectively.

In this section, we are interested in designing a full-order switched filter described by

$$\begin{aligned} \hat{x}_{k+1} &= A_{f\sigma} \hat{x}_k + B_{f\sigma} y_k, \\ \hat{z}_k &= C_{f\sigma} \hat{x}_k + D_{f\sigma} y_k, \end{aligned} \tag{4.2}$$

where the matrices A_{fi}, B_{fi}, C_{fi} and D_{fi}, $i \in \mathscr{P}$ are the filter parameters to be determined. The filter (4.2) is assumed to be switched synchronously by the switching signal σ in system (4.1).

Augmenting the model of system (4.1) to include the states of (4.2), we can obtain the following error system:

$$\begin{aligned} \tilde{x}_{k+1} &= \tilde{A}_\sigma \tilde{x}_k + \tilde{A}_{d\sigma} E \tilde{x}_{k-d_k} + \tilde{B}_\sigma w_k, \\ \tilde{z}_k &= \tilde{C}_\sigma \tilde{x}_k + \tilde{C}_{d\sigma} E \tilde{x}_{k-d_k} + \tilde{D}_\sigma w_k, \end{aligned} \tag{4.3}$$

where

$$\begin{aligned} \tilde{x}_k &= \begin{bmatrix} x_k \\ \hat{x}_k \end{bmatrix}, \\ \tilde{z}_k &= z_k - \hat{z}_k, E = \begin{bmatrix} I & 0 \end{bmatrix}, \\ \tilde{A}_i &= \begin{bmatrix} A_i & 0 \\ B_{fi} C_i & A_{fi} \end{bmatrix}, \\ \tilde{A}_{di} &= \begin{bmatrix} A_{di} \\ B_{fi} C_{di} \end{bmatrix}, \\ \tilde{B}_i &= \begin{bmatrix} B_i \\ B_{fi} D_i \end{bmatrix}, \\ \tilde{C}_i &= \begin{bmatrix} G_i - D_{fi} C_i & -C_{fi} \end{bmatrix}, \\ \tilde{C}_{di} &= G_{di} - D_{fi} C_{di}, \\ \tilde{D}_i &= E_i - D_{fi} D_i. \end{aligned}$$

Before presenting the main results, the following definitions are given.

Definition 4.1. *([42]) Construct a finite switching sequence σ during $[k_0, k)$, where switching time instants $k_0 < k_1 < \cdots < k_{s-1} < k_s, s = 1, 2, \cdots$. Then, for any $k_v > k_s > 0$, let $N_\sigma(k_s, k_v)$ denote the switching numbers of σ during $[k_s, k_v)$. For given scalars $\tau^* > 0$ and $N_0 \geq 0$, we have*

$$N_\sigma(k_s,\, k_v) \leq N_0 + \frac{k_v - k_s}{\tau^*}, \tag{4.4}$$

where τ^* and N_0 are called ADT and the chattering bound, respectively. Here we assume $N_0 = 0$ as commonly used in literature.

Definition 4.2. *([131]) The exponential H_∞ filtering: given $\gamma > 0$, the filtering error systems are exponentially stable with an H_∞ norm bound γ if they are exponentially stable and, under zero-initial condition, the induced l_2-norm of the operator from the noise signals w_k to the estimation error $\tilde{z}_k$ is less than γ, i.e., $\|\tilde{z}_k\|_2 \leq \gamma \|w_k\|_2$, for all nonzero $w_k \in l_2[0, \infty)$.*

The exponential H_∞ filtering problem to be addressed in this paper can be formulated as follows: given system (4.1) and an H_∞ norm bound γ, develop a suitable filter in the form of (4.2), and find admissible ADT switching signals such that the following specifications are met:

(1) System (4.3) is exponentially stable when $w_k = 0$;
(2) System (4.3) guarantees, under zero-initial condition, $\|\tilde{z}_k\|_2 \leq \gamma \|w_k\|_2$ for all nonzero $w_k \in l_2[0,\infty)$.

4.1.2 Stability and H_∞ Performance Analysis

This section gives a new characterization involving ADT method and Lyapunov-Krasovskii function approach for system (4.3). First, exponential H_∞ performance analysis is presented as follows.

Theorem 4.3. *Given scalars $0 < \alpha < 1$, $\gamma > 0$, $\mu \geq 1$ and the switching time instants $k_0 < k_1 < \cdots < k_{s-1} < k_s, s = 1,2,\cdots$, during $[k_0,k)$, the filtering error system (4.3) is exponentially stable with an H_∞ norm bound γ under ADT switching signal σ_{k_s} (denoting $\sigma(k_s)$ for simplicity), if there exist $V_{\sigma_{k_s}}(k)$ (denoting $V_{\sigma_{k_s}}(\tilde{x}_k)$ for simplicity) satisfying*

$$V_{\sigma_{k_s}}(k) \leq \alpha^{k-k_s} V_{\sigma_{k_s}}(k_s) - \sum_{l=k_s}^{k-1} \alpha^{k-l-1} \Gamma(l), \tag{4.5}$$

$$V_{\sigma_{k_s}}(k) \leq \mu V_{\sigma_{k_{s-1}}}(k), \tag{4.6}$$

and the average dwell time τ_a

$$\tau_a > \tau^* = -\ln\mu / \ln\alpha, \tag{4.7}$$

where $\Gamma(l) = \tilde{z}_l^T \tilde{z}_l - \gamma^2 w_l^T w_l$.

Proof. By combining (4.5) with (4.6), one can get

$$\begin{aligned}
V_{\sigma_{k_s}}(k) &\leq \alpha^{k-k_s} V_{\sigma_{k_s}}(k_s) - \sum_{l=k_s}^{k-1} \alpha^{k-l-1} \Gamma(l) \\
&\leq \alpha^{k-k_s} \mu V_{\sigma_{k_{s-1}}}(k_s) - \sum_{l=k_s}^{k-1} \alpha^{k-l-1} \Gamma(l) \\
&\leq \alpha^{k-k_{s-1}} \mu V_{\sigma_{k_{s-1}}}(k_{s-1}) - \mu \sum_{l=k_{s-1}}^{k_s-1} \alpha^{k-l-1} \Gamma(l) - \sum_{l=k_s}^{k-1} \alpha^{k-l-1} \Gamma(l) \\
&\leq \alpha^{k-k_0} \mu^{N_\sigma(k_0,k)} V_{\sigma_{k_0}}(k_0) - \mu^{N(k_0,k)} \sum_{l=k_0}^{k_1-1} \alpha^{k-l-1} \Gamma(l) \\
&\quad -\mu^{N(k_1,k)} \sum_{l=k_1}^{k_2-1} \alpha^{k-l-1} \Gamma(l) - \cdots - \sum_{l=k_s}^{k-1} \alpha^{k-l-1} \Gamma(l) \\
&\leq \alpha^{k-k_0} \mu^{N_\sigma(k_0,k)} V_{\sigma_{k_0}}(k_0) - \sum_{l=k_0}^{k-1} \mu^{N_\sigma(l,k)} \alpha^{k-l-1} \Gamma(l).
\end{aligned} \tag{4.8}$$

Considering (4.4), (4.7) and (4.8), by using ADT technique, we have

$$V_{\sigma_{k_s}}(k) \leq e^{\lambda(k-k_0)} V_{\sigma_{k_0}}(k_0) - \sum_{l=k_0}^{k-1} e^{\lambda(k-l)} \alpha^{-1} \Gamma(l), \tag{4.9}$$

where $\lambda = \ln\alpha + \ln\mu/\tau_a$. It follows from (4.7) that $\lambda < 0$.

Assuming the zero disturbance input $w_k = 0$ to the state equation of the filtering error system (4.3), it follows from (4.9) that

$$V_{\sigma_{k_s}}(k) \leq e^{\lambda(k-k_0)} V_{\sigma_{k_0}}(k_0).$$

In addition, there exist some constants $a_{\sigma_{k_s}} > 0$ and $b_{\sigma_{k_0}} > 0$ such that

$$a_{\sigma_{k_s}} \left\|\tilde{x}_k\right\|^2 \leq V_{\sigma_{k_s}}(k),$$

$$V_{\sigma_{k_0}}(k_0) \leq b_{\sigma_{k_0}} \left\|\tilde{x}_{k_0}\right\|^2.$$

Then, we have

$$\left\|\tilde{x}_k\right\| \leq \sqrt{b_{\sigma_{k_0}}/a_{\sigma_{k_s}}}\, e^{\lambda(k-k_0)/2} \left\|\tilde{x}_{k_0}\right\|.$$

Therefore, the filtering error system (4.3) is exponentially stable.

Now, we consider the following performance index:

$$J = \sum_{k=0}^{\infty} \left[\tilde{z}_k^T \tilde{z}_k - \gamma^2 w_k^T w_k\right].$$

For any nonzero $w_k \in l_2[0,\infty)$ and under zero-initial condition, from (4.9), one has

$$\sum_{l=0}^{k-1} e^{\lambda(k-l)} \alpha^{-1} \Gamma(l) \leq 0.$$

Note in the above inequality that

$$\sum_{l=0}^{k-1} e^{\lambda(k-l)} \alpha^{-1} \tilde{z}_l^T \tilde{z}_l,$$

is summable from $k = 1$ to ∞ since

$$\sum_{l=0}^{k-1} e^{\lambda(k-l)} \alpha^{-1} w_l^T w_l,$$

is summable for any $w_l \in l_2[0,\infty)$ in the same interval. Then, from $k = 1$ to ∞, we obtain that

$$\sum_{k=1}^{\infty} \sum_{l=0}^{k-1} e^{\lambda(k-l)} \alpha^{-1} \Gamma(l) \leq 0.$$

Rearranging the double-sum area yields

$$\sum_{l=0}^{\infty}\Gamma(l)\sum_{k=l+1}^{\infty}e^{\lambda(k-l)}\alpha^{-1}=\frac{e^{\lambda}\alpha^{-1}}{1-e^{\lambda}}\sum_{l=0}^{\infty}\Gamma(l)\leq 0,$$

which means that $J\leq 0$. Then, one has $\|\tilde{z}_k\|_2\leq\gamma\|w_k\|_2$ for nonzero $w_k\in l_2[0,\infty)$. The proof is completed.

Remark 4.4. It is well known that two conditions are to be satisfied for analysis and synthesis of switched systems under ADT switching signals [138, 141, 120]. First, Lyapunov function values of each subsystem exponentially decay. Second, Lyapunov function values are considered to increase μ times ($\mu\geq 1$) at switching time instants, which is required for the exponential stability of switched systems. In this paper, they are formulated as (4.5) and (4.6), respectively. Therefore, exponential H_∞ performance criteria are introduced to make such filter design feasible.

Now, we are to present two inequalities for system (4.3) satisfying (4.5) and (4.6).

Theorem 4.5. *For system (4.3), $0<\alpha<1$, $\gamma>0$ and $\mu\geq 1$, (4.5) and (4.6) hold if there exist symmetric and positive definite matrices P_i, Q_i, R_{bi}, Z_{bi} and matrices Ω_i, $\hat{\Omega}_i$, M_{ci}, S_{ci}, N_{ci} with $b=1,2,c=1,2,3,\forall i\in\mathscr{P}$, such that the following inequalities hold,*

$$\begin{bmatrix}\Xi_1 & \Xi_2\\ * & \Xi_3\end{bmatrix}<0, \tag{4.10}$$

$$\begin{aligned}&P_i\leq\mu P_j,\ Q_i\leq\mu Q_j,\\ &R_{bi}\leq\mu R_{bj},\ Z_{bi}\leq\mu Z_{bj},\ \forall i,\ j\in\mathscr{P},\end{aligned} \tag{4.11}$$

where

$$\begin{aligned}
\Xi_1&=\begin{bmatrix}\Psi_{11} & \Psi_{12} & \Psi_{13}\\ * & \Psi_{22} & \Psi_{23}\\ * & * & -\gamma^2 I\end{bmatrix},\\
\Xi_2&=\begin{bmatrix}N_{1i} & -S_{1i} & M_{1i} & S_{1i} & N_{1i} & \tilde{C}_i^T & \tilde{A}_i^T\Omega_i^T & \Psi_{111}\hat{\Omega}_i^T\\ N_{2i} & -S_{2i} & M_{2i} & S_{2i} & N_{2i} & \tilde{C}_{di}^T & \tilde{A}_{di}^T\Omega_i^T & A_{di}^T\hat{\Omega}_i^T\\ N_{3i} & -S_{3i} & M_{3i} & S_{3i} & N_{3i} & \tilde{D}_i^T & \tilde{B}_i^T\Omega_i^T & B_i^T\hat{\Omega}_i^T\end{bmatrix},\\
\Xi_3&=diag\left\{\Psi_{44},\Psi_{55},-Z_{1i},-Z_{3i},-Z_{2i},-I,\Psi_{1010},\Psi_{1111}\right\},\\
\Psi_{1010}&=P_i-\left(\Omega_i+\Omega_i^T\right),\\
\Psi_{11}&=E^T(\tau Q_i+R_{1i}+R_{2i})E-\alpha P_i+E^TM_{1i}^T+M_{1i}E,\\
\Psi_{12}&=E^TM_{2i}^T+S_{1i}-N_{1i}-M_{1i},\\
\Psi_{22}&=S_{2i}^T-\alpha^{d_M}Q_i+S_{2i}-M_{2i}^T-M_{2i}-N_{2i}^T-N_{2i},\\
\Psi_{23}&=S_{3i}^T-M_{3i}^T-N_{3i}^T,\\
\Psi_{44}&=-\alpha^{d_m}R_{1i},\ \Psi_{55}=-\alpha^{d_M}R_{2i},\\
\Psi_{111}&=E^T(A_i^T-I),\ \Psi_{1111}=Z_{4i}-\left(\hat{\Omega}_i+\hat{\Omega}_i^T\right),\\
\Psi_{13}&=E^TM_{3i}^T,\ Z_{3i}=Z_{1i}+Z_{2i},\\
Z_{4i}&=d_MZ_{1i}+\tau Z_{2i},\ \tau=d_M-d_m+1.
\end{aligned}$$

Proof. Choosing the following Lyapunov-Krasovskii function as

$$V_i(k) = V_{1i}(k) + V_{2i}(k) + V_{3i}(k) + V_{4i}(k) + V_{5i}(k) + V_{6i}(k), \tag{4.12}$$

where

$$\begin{aligned}
V_{1i}(k) &= \tilde{x}_k^T P_i \tilde{x}_k, \\
V_{2i}(k) &= \sum_{l=k-d_k}^{k-1} \alpha^{k-l-1} x_l^T Q_i x_l + \sum_{j=-d_M+1}^{-d_m} \sum_{l=k+j}^{k-1} \alpha^{k-l-1} x_l^T Q_i x_l, \\
V_{3i}(k) &= \sum_{l=k-d_m}^{k-1} \alpha^{k-l-1} x_l^T R_{1i} x_l, \\
V_{4i}(k) &= \sum_{l=k-d_M}^{k-1} \alpha^{k-l-1} x_l^T R_{2i} x_l, \\
V_{5i}(k) &= \sum_{j=-d_M}^{-1} \sum_{l=k+j}^{k-1} \alpha^{k-l-1} \eta_l^T Z_{1i} \eta_l, \\
V_{6i}(k) &= \sum_{j=-d_M}^{-d_m} \sum_{l=k+j}^{k-1} \alpha^{k-l-1} \eta_l^T Z_{2i} \eta_l, \\
\eta_l &= x_{l+1} - x_l,
\end{aligned}$$

and $P_i, Q_i, R_{1i}, R_{2i}, Z_{1i}, Z_{2i}$ are symmetric and positive definite matrices.
In terms of $\Delta V_i(k) = V_i(k+1) - \alpha V_i(k)$, we have

$$\Delta V_{1i} = \tilde{x}_{k+1}^T P_i \tilde{x}_{k+1} - \alpha \tilde{x}_k^T P_i \tilde{x}_k, \tag{4.13}$$

Note that

$$\begin{aligned}
&\sum_{l=k+1-d_{k+1}}^{k-1} \alpha^{k-l} x_l^T Q_i x_l \\
&\le \sum_{l=k+1-d_k}^{k-1} \alpha^{k-l} x_l^T Q_i x_l + \sum_{l=k+1-d_M}^{k-d_m} \alpha^{k-l} x_l^T Q_i x_l.
\end{aligned}$$

Thus, we have

$$\begin{aligned}
\Delta V_{2i} &= \tau x_k^T Q_i x_k - \alpha^{d_k} x_{k-d_k}^T Q_i x_{k-d_k} \\
&\quad + \sum_{l=k+1-d_{k+1}}^{k-1} \alpha^{k-l} x_l^T Q_i x_l - \sum_{l=k+1-d_k}^{k-1} \alpha^{k-l} x_l^T Q_i x_l \\
&\quad - \sum_{l=k+1-d_M}^{k-d_m} \alpha^{k-l} x_l^T Q_i x_l \\
&\le \tau x_k^T Q_i x_k - \alpha^{d_M} x_{k-d_k}^T Q_i x_{k-d_k}.
\end{aligned} \tag{4.14}$$

Similarly, we get

$$\Delta V_{3i} = x_k^T R_{1i} x_k - \alpha^{d_m} x_{k-d_m}^T R_{1i} x_{k-d_m}, \tag{4.15}$$

$$\Delta V_{4i} = x_k^T R_{2i} x_k - \alpha^{d_M} x_{k-d_M}^T R_{2i} x_{k-d_M}, \tag{4.16}$$

$$\begin{aligned}\Delta V_{5i} =& d_M \eta_k^T Z_{1i} \eta_k \\ &- \sum_{l=k-d_k}^{k-1} \alpha^{k-l} \eta_l^T Z_{1i} \eta_l - \sum_{l=k-d_M}^{k-d_k-1} \alpha^{k-l} \eta_l^T Z_{1i} \eta_l,\end{aligned} \tag{4.17}$$

$$\begin{aligned}\Delta V_{6i} =& (d_M - d_m + 1) \eta_k^T Z_{2i} \eta_k - \sum_{l=k-d_k}^{k-d_m} \alpha^{k-l} \eta_l^T Z_{2i} \eta_l \\ &- \sum_{l=k-d_M}^{k-d_k-1} \alpha^{k-l} \eta_l^T Z_{2i} \eta_l.\end{aligned} \tag{4.18}$$

Defining

$$\xi_k = \left[\tilde{x}_k^T \; x_{k-d_k}^T \; w_k^T \; x_{k-d_m}^T \; x_{k-d_M}^T \right]^T,$$

one obtains

$$\begin{aligned}
&\Delta V_i(k) + \tilde{z}_k^T \tilde{z}_k - \gamma^2 w_k^T w_k \\
&= \Delta V_i(k) + \tilde{z}_k^T \tilde{z}_k - \gamma^2 w_k^T w_k \\
&+2\xi_k^T M_i \left[x_k - x_{k-d_k} - \sum_{l=k-d_k}^{k-1} \eta_l \right] \\
&+2\xi_k^T S_i \left[x_{k-d_k} - x_{k-d_M} - \sum_{l=k-d_M}^{k-d_k-1} \eta_l \right] \\
&+2\xi_k^T N_i \left[x_{k-d_m} - x_{k-d_k} - \sum_{l=k-d_k}^{k-d_m-1} \eta_l \right] \\
&\leq \xi_k^T \left(\Phi_i + M_i Z_{1i}^{-1} M_i^T + S_i Z_{3i}^{-1} S_i^T + N_i Z_{2i}^{-1} N_i^T \right) \xi_k \\
&+\tilde{z}_k^T \tilde{z}_k + \tilde{x}_{k+1}^T P_i \tilde{x}_{k+1} + \eta_k^T Z_{4i} \eta_k \\
&- \sum_{l=k-d_k}^{k-1} \left[\xi_k^T M_i + \alpha^{k-l} \eta_l^T Z_{1i} \right] \alpha^{l-k} Z_{1i}^{-1} \left[M_i^T \xi_k + \alpha^{k-l} Z_{1i} \eta_l \right] \\
&- \sum_{l=k-d_M}^{k-d_k-1} \left[\xi_k^T S_i + \alpha^{k-l} \eta_l^T Z_{3i} \right] \alpha^{l-k} Z_{3i}^{-1} \left[S_i^T \xi_k + \alpha^{k-l} Z_{3i} \eta_l \right] \\
&- \sum_{l=k-d_k}^{k-d_m} \left[\xi_k^T N_i + \alpha^{k-l} \eta_l^T Z_{2i} \right] \alpha^{l-k} Z_{2i}^{-1} \left[N_i^T \xi_k + \alpha^{k-l} Z_{2i} \eta_l \right],
\end{aligned} \tag{4.19}$$

where

$$\Phi_i = \begin{bmatrix} \Psi_{11} & \Psi_{12} & \Psi_{13} & N_{1i} & -S_{1i} \\ * & \Psi_{22} & \Psi_{23} & N_{2i} & -S_{2i} \\ * & * & -\gamma^2 I & N_{3i} & -S_{3i} \\ * & * & * & \Psi_{44} & 0 \\ * & * & * & * & \Psi_{55} \end{bmatrix}.$$

Since $Z_{1i} > 0$ and $Z_{2i} > 0$, the last three terms are all nonpositive. By Schur Complement, we have

$$\Delta V_i(k) + \tilde{z}_k^T \tilde{z}_k - \gamma^2 w_k^T w_k \leq 0, \tag{4.20}$$

if the following inequality holds,

$$\begin{bmatrix} \Xi_1 & \Xi_4 \\ * & \Xi_5 \end{bmatrix} < 0, \tag{4.21}$$

where

$$\Xi_4 = \begin{bmatrix} N_{1i} & -S_{1i} & M_{1i} & S_{1i} & N_{1i} & \tilde{C}_i^T & \tilde{A}_i^T & \Psi_{111} \\ N_{2i} & -S_{2i} & M_{2i} & S_{2i} & N_{2i} & \tilde{C}_{di}^T & \tilde{A}_{di}^T & A_{di}^T \\ N_{3i} & -S_{3i} & M_{3i} & S_{3i} & N_{3i} & \tilde{D}_i^T & \tilde{B}_i^T & B_i^T \end{bmatrix},$$

$$\Xi_5 = diag\left\{ \Psi_{44}, \Psi_{55}, -Z_{1i}, -Z_{3i}, -Z_{2i}, -I, -P_i^{-1}, -Z_{4i}^{-1} \right\}.$$

Now, it is readily seen from (4.10) that

$$P_i - \left(\Omega_i^T + \Omega_i\right) < 0,$$

which means that Ω_i is nonsingular. Meanwhile, $P_i > 0$, and from the fact that

$$(P_i - \Omega_i) P_i^{-1} (P_i - \Omega_i)^T \geq 0,$$

we have

$$-\Omega_i P_i^{-1} \Omega_i^T \leq P_i - \left(\Omega_i + \Omega_i^T\right).$$

Similarly, one has

$$-\hat{\Omega}_i Z_{4i}^{-1} \hat{\Omega}_i^T \leq Z_{4i} - \left(\hat{\Omega}_i + \hat{\Omega}_i^T\right).$$

Therefore, from (4.10), we can conclude that

$$\begin{bmatrix} \Xi_1 & \Xi_2 \\ * & \Xi_6 \end{bmatrix} < 0, \tag{4.22}$$

where

$$\Xi_6 = diag\left\{ \Psi_{44}, \Psi_{55}, -Z_{1i}, -Z_{3i}, -Z_{2i}, -I, \hat{\Psi}_{1010}, \hat{\Psi}_{1111} \right\},$$
$$\hat{\Psi}_{1010} = -\Omega_i P_i^{-1} \Omega_i^T,$$
$$\hat{\Psi}_{1111} = -\hat{\Omega}_i Z_{4i}^{-1} \hat{\Omega}_i^T.$$

Performing a congruence transformation to (4.22) via

$$diag\left\{I,I,I,I,I,I,I,I,I,\Omega_i^{-T},\hat{\Omega}_i^{-T}\right\},$$

yields (4.21). This means that if (4.10) holds, (4.20) is true. Furthermore, one can obtain from (4.20) that

$$V_i(k+1) \leq \alpha V_i(k) - \Gamma(k). \tag{4.23}$$

During $[k_s,k)$, iterating (4.23) leads to (4.5). On the other hand, it follows from (4.11) and (4.12) that $V_i(k) \leq \mu V_j(k)$, which implies that (4.6) holds. This proof is completed.

Remark 4.6. Note that Theorem 4.5 presents (4.10) and (4.11) such that the Lyapunov-Krasovskii function candidate (4.12) satisfied (4.5) and (4.6). While (4.6) is easily satisfied, the main difficulty lies in (4.5). For switched delay-free systems, it is also easy to find the condition satisfying (4.5). For example, in [141],

$$V_i(\tilde{x}_k) = \tilde{x}_k^T P_i \tilde{x}_k,$$

is constructed to find a condition satisfying

$$V_i(\tilde{x}_{k+1}) - \alpha V_i(\tilde{x}_k) < 0,$$

with $0 < \alpha < 1$. In such a condition, α is involved. However, for switched delay systems, it is obviously difficult to establish the conditions satisfying

$$V_i(\tilde{x}_{k+1}) - \alpha V_i(\tilde{x}_k) < 0,$$

with $0 < \alpha < 1$ due to the limitation of classic Lyapunov-Krasovskii functions. For example, in [89], it is observed from $V_i(\tilde{x}_{k+1})$ that

$$\begin{aligned}
&V_{3i}(k+1) - \alpha V_{3i}(k) \\
&= \sum_{l=k+1-d_m}^{k} x_l^T R_{1i} x_l - \sum_{l=k-d_m}^{k-1} \alpha x_l^T R_{1i} x_l \\
&= x_k^T R_{1i} x_k + \sum_{l=k+1-d_m}^{k-1} x_l^T (1-\alpha) R_{1i} x_l - \alpha x_{k-d_m}^T R_{1i} x_{k-d_m}.
\end{aligned}$$

Due to the existence of the summation,

$$\sum_{l=k+1-d_m}^{k-1} x_l^T (1-\alpha) R_{1i} x_l,$$

it is difficult to establish a condition satisfying

$$V_{3i}(k+1) - \alpha V_{3i}(k) < 0.$$

However, the difficulty is overcome by introducing the factor α^{k-l-1} in (4.12), i.e.,

$$V_{3i}(k) = \sum_{l=k-d_m}^{k-1} \alpha^{k-l-1} x_l^T R_{1i} x_l.$$

By using $\alpha V_{3i}(k)$, it is seen from (4.15) that

$$\begin{aligned} & V_{3i}(k+1) - \alpha V_{3i}(k) \\ = & \sum_{l=k+1-d_m}^{k} \alpha^{k-l} x_l^T R_{1i} x_l - \sum_{l=k-d_m}^{k-1} \alpha^{k-l} x_l^T R_{1i} x_l \\ = & \, x_k^T R_{1i} x_k - \alpha^{d_m} x_{k-d_m}^T R_{1i} x_{k-d_m}, \end{aligned}$$

which makes the condition obtained easily. The same methods are adopted in (4.12) except for $V_{1i}(k)$. This enables us to give the condition satisfying (4.5). Thus, the proposed method can make ADT method applicable and has the potential to yield the results for other problems of switched delay systems under ADT switching signals.

Remark 4.7. It is also noted that the proposed Lyapunov-Krasovskii function (4.12) for H_∞ performance analysis is more general than those in literature[89, 15, 72, 141, 144, 146].

1. When $\alpha = 0$, (4.12) is reduced to quadratic Lyapunov function

$$V_i(\tilde{x}_k) = \tilde{x}_k^T P_i \tilde{x}_k, \tag{4.24}$$

which is the same as that in [141, 146]. Based on (4.24) and SLF method, H_∞ filters for switched systems are designed in [146]. Also, based on (4.24) and ADT method, H_∞ filters for switched systems are constructed in [141].

2. When $\alpha = 1$, (4.12) becomes the classic Lyapunov-Krasovskii function:

$$\begin{aligned} V_i(k) = & \, \tilde{x}_k^T P_i \tilde{x}_k + \sum_{l=k-d_k}^{k-1} x_l^T Q_i x_l + \sum_{j=-d_M+1}^{-d_m} \sum_{l=k+j}^{k-1} x_l^T Q_i x_l \\ & + \sum_{l=k-d_m}^{k-1} x_l^T R_{1i} x_l + \sum_{l=k-d_M}^{k-1} x_l^T R_{2i} x_l \\ & + \sum_{j=-d_M}^{-1} \sum_{l=k+j}^{k-1} \eta_l^T Z_{1i} \eta_l + \sum_{j=-d_M}^{-d_m} \sum_{l=k+j}^{k-1} \eta_l^T Z_{2i} \eta_l. \end{aligned} \tag{4.25}$$

For non-switched delay systems, the results in [89] can be obtained by means of (4.25) combined with Finsler's lemma. For switched systems with constant delays, i.e. $d_k \equiv d$, based on SLF method, the first two terms of (4.25) are exploited to present delay-independent conditions on the existence of H_∞ filter in [15]. For switched systems with interval time-varying delays, based on SLF method and (4.25) with common Lyapunov matrices $Q_i \equiv Q, R_{bi} \equiv R_b, Z_{bi} \equiv Z_b, b = 1,2$, H_∞ and L_2-L_∞ performance analysis are established in [72] and [144], respectively.

It is shown that Lyapunov-Krasovskii function (4.12) can cover quadratic Lyapunov function (4.24) and the classic Lyapunov-Krasovskii function (4.25) in literature.

Remark 4.8. ADT method is less conservative than SLF method when they are used to analyze and synthesize the performance of switched delay systems[144]. In Remark 4 of [144], it is pointed out that only P_i is switched Lyapunov matrix while Q and Z are common Lyapunov matrices and it is very difficult to choose Q_i and Z_i due to the tight coupling between them and the time-delay terms. There exists the same conservatism in [72]. The difficulty is overcome by introducing α and thus the proposed Lyapunov-Krasovskii function (4.12) does not include any common Lyapunov matrix. All Lyapunov matrices are chosen as piecewise matrices such as P_i, Q_i, R_{bi}, Z_{bi}, $b=1,2$. From this point of view, our results are also less conservative.

4.1.3 Filter Design

In this section, we will obtain sufficient conditions for the existence of a filter in form of (4.2) assuring an exponential H_∞ performance for system (4.1).

Theorem 4.9. *Consider system (4.1) and let $0<\alpha<1$, $\gamma>0$ and $\mu\geq 1$ be given constants, if there exist symmetric and positive definite matrices P_{bi}, Q_i, R_{bi}, Z_{bi}, and matrices P_{3i}, $\tilde{M}_{1i}$, $\tilde{S}_{1i}$, $\tilde{N}_{1i}$, $M_{(b+1)i}$, $S_{(b+1)i}$, $N_{(b+1)i}$, Z_i, $\bar{M}_i$, $\hat{\Omega}_i$, L_i, H_i, $\bar{S}_i$, $\bar{N}_i$ with $b=1,2$, $i\in\mathscr{P}$, such that the following inequalities hold,*

$$\begin{bmatrix}\tilde{\Xi}_1 & \tilde{\Xi}_4\\ * & \tilde{\Xi}_5\end{bmatrix}<0, \tag{4.26}$$

$$\tilde{P}_i\leq\mu\tilde{P}_j, Q_i\leq\mu Q_j, R_{bi}\leq\mu R_{bj}, Z_{bi}\leq\mu Z_{bj}, \{i,\ j\}\in\mathscr{P}, \tag{4.27}$$

where

$$\tilde{\Xi}_1=\begin{bmatrix}\tilde{\Psi}_{11} & \tilde{\Psi}_{12} & \Psi_{13}\\ * & \Psi_{22} & \Psi_{23}\\ * & * & -\gamma^2 I\end{bmatrix},$$

$$\tilde{\Xi}_4=\begin{bmatrix}\tilde{N}_{1i} & -\tilde{S}_{1i} & \tilde{M}_{1i} & \tilde{S}_{1i} & \tilde{N}_{1i} & \tilde{\Psi}_{19}^T & \tilde{\Psi}_{110}^T & \Psi_{111}\hat{\Omega}_i^T\\ N_{2i} & -S_{2i} & M_{2i} & S_{2i} & N_{2i} & \tilde{\Psi}_{29}^T & \tilde{\Psi}_{210}^T & A_{di}^T\hat{\Omega}_i^T\\ N_{3i} & -S_{3i} & M_{3i} & S_{3i} & N_{3i} & \tilde{\Psi}_{39}^T & \tilde{\Psi}_{310}^T & B_i^T\hat{\Omega}_i^T\end{bmatrix},$$

$$\tilde{\Xi}_5=diag\left\{\Psi_{44},\Psi_{55},-Z_{1i},-Z_{3i},-Z_{2i},-I,\tilde{\Psi}_{1010},\Psi_{1111}\right\},$$

$$\tilde{\Psi}_{11}=-\alpha\tilde{P}_i+E^T(\tau Q_i+R_{1i}+R_{2i})E+E^T\tilde{M}_{1i}^T+\tilde{M}_{1i}E,$$

$$\tilde{\Psi}_{12}=E^TM_{2i}^T+\tilde{S}_{1i}-\tilde{N}_{1i}-\tilde{M}_{1i},$$

$$\tilde{\Psi}_{19}=\left[G_i-\bar{N}_iC_i\ -\bar{S}_i\right],$$

$$\tilde{\Psi}_{29}=G_{di}-\bar{N}_iC_{di},$$

$$\tilde{\Psi}_{39}=E_i-\bar{N}_iD_i,$$

$$\tilde{\Psi}_{110}=\begin{bmatrix}Z_iA_i+H_iC_i & L_i\\ H_iC_i & L_i\end{bmatrix},$$

$$\tilde{\Psi}_{210}=\begin{bmatrix}Z_iA_{di}+H_iC_{di}\\ H_iC_{di}\end{bmatrix},$$

$$\tilde{\Psi}_{310}=\begin{bmatrix}Z_iB_i+H_iD_i\\ H_iD_i\end{bmatrix},$$

$$\tilde{\Psi}_{1010} = \tilde{P}_i - \left(\tilde{\Omega}_i + \tilde{\Omega}_i^T\right),$$
$$\tilde{\Omega}_i = \begin{bmatrix} Z_i & \bar{M}_i \\ 0 & \bar{M}_i \end{bmatrix}, \ \tilde{P}_i = \begin{bmatrix} P_{1i} & P_{3i} \\ * & P_{2i} \end{bmatrix}.$$

Then, there exists a filter in the form of (4.2) such that the filtering error system (4.3) is exponentially stable with an H_∞ norm bound γ under ADT switching signals satisfying (4.7). Moreover, if (4.26) and (4.27) have a feasible solution, the parameters of an admissible filter can be constructed by

$$\begin{bmatrix} A_{fi} & B_{fi} \\ C_{fi} & D_{fi} \end{bmatrix} = \begin{bmatrix} \hat{U}_i^{-1} L_i \bar{M}_i^{-1} \hat{U}_i & \hat{U}_i^{-1} H_i \\ \bar{S}_i \bar{M}_i^{-1} \hat{U}_i & \bar{N}_i \end{bmatrix}, \tag{4.28}$$

where $\hat{U}_i \in R^{n \times n}$ is any invertible matrix (for example, $\hat{U}_i$ could be set as I).

Proof. By Theorems 4.3 and 4.5, the filtering error system (4.3) under ADT switching signals satisfying (4.7) is exponentially stable with an H_∞ norm bound γ, if LMIs (4.11) and (4.10) hold.

Note that from (4.26), we have

$$\begin{bmatrix} Z_i + Z_i^T & \bar{M}_i \\ * & \bar{M}_i + \bar{M}_i^T \end{bmatrix} > 0, \tag{4.29}$$

which means that Z_i and $\bar{M}_i$ are nonsingular. And thus, if (4.26) holds, there exist nonsingular matrices $\hat{U}_i$ and U_i satisfying $\bar{M}_i = \hat{U}_i U_i^{-T} \hat{U}_i^T$.

Now, we introduce

$$F_i = \begin{bmatrix} I & 0 \\ 0 & \hat{U}_i U_i^{-1} \end{bmatrix}, \ \Omega_i = \begin{bmatrix} Z_i & \hat{U}_i \\ 0 & U_i \end{bmatrix},$$
$$H_i \triangleq \hat{U}_i B_{fi}, \ L_i \triangleq \hat{U}_i A_{fi} U_i^{-T} \hat{U}_i^T,$$
$$\bar{S}_i \triangleq C_{fi} U_i^{-T} \hat{U}_i^T, \ \bar{N}_i \triangleq D_{fi},$$
$$\tilde{M}_{1i} \triangleq F_i M_{1i}, \ \tilde{S}_{1i} \triangleq F_i S_{1i},$$
$$\tilde{N}_{1i} \triangleq F_i N_{1i}, \ \tilde{P}_i \triangleq F_i P_i F_i^T. \tag{4.30}$$

By using (4.3), (4.26) and (4.30), we can get

$$F_i \Psi_{11} F_i^T = \tilde{\Psi}_{11}, \ F_i \Psi_{12} = \tilde{\Psi}_{12},$$
$$F_i \Omega_i \tilde{A}_i F_i^T = \tilde{\Psi}_{110}, \ F_i \Omega_i \tilde{A}_{di} = \tilde{\Psi}_{210},$$
$$F_i \Omega_i \tilde{B}_i = \tilde{\Psi}_{310}, \ \tilde{C}_i F_i^T = \tilde{\Psi}_{19}. \tag{4.31}$$

By Schur Complement, performing a congruence transformation to (4.10) via

$$diag\left\{F_i^T, I, I, I, I, I, I, I, I, F_i^T, I\right\},$$

yields (4.26), which implies that (4.10) holds. Meanwhile, by means of $\tilde{P}_i = F_i P_i F_i^T$ in (4.30), it is obtained that if (4.27) holds, then (4.11) is true. Therefore, we can conclude that the filter parameters defined in (4.30) guarantees the filtering error system (4.3) to be exponentially stable with an H_∞ norm bound γ. In addition, we know from (4.30) that the parameters of an admissible filter are given by

$$\begin{bmatrix} A_{fi} & B_{fi} \\ C_{fi} & D_{fi} \end{bmatrix} = \begin{bmatrix} \hat{U}_i^{-1} L_i \hat{U}_i^{-T} U_i^T & \hat{U}_i^{-1} H_i \\ \bar{S}_i \hat{U}_i^{-T} U_i^T & \bar{N}_i \end{bmatrix}. \tag{4.32}$$

It should be pointed out that the filter parameters A_{fi}, B_{fi} and C_{fi} can not be directly obtained from (4.32) except D_{fi} because $\hat{U}_i$ and U_i can not be derived from the solutions of LMI (4.26) but $\bar{N}_i$. Noting that $\bar{M}_i = \hat{U}_i U_i^{-T} \hat{U}_i^T$, we have $\bar{M}_i^{-1}\hat{U}_i = \hat{U}_i^{-T} U_i^T$. In addition, $\bar{M}_i$ can be solved from the solutions of LMI (4.26) and $\hat{U}_i$ can be chosen as any invertible matrix. Substituting $\bar{M}_i^{-1}\hat{U}_i$ into (4.32) yields (4.28) which means that the parameters of an admissible filter described by (4.28) can be given by solving LMIs (4.26) and (4.27). The proof is completed.

Remark 4.10. From (4.9), it is observed that α is associated with the decay rate of exponential stability. The bigger it is, the smaller the decay rate is. Therefore, in general, by choosing a large α it is easier to find a feasible solution.

Design Algorithm:
Step 1: In light of Remark 4.10, we choose a large α in $(0,\ 1)$. However, μ is selected to a small one in $(1,\ \infty)$. Then, by Theorem 3, given known constants α and μ, which are close to one, solve LMIs (4.7), (4.26) and (4.27).
Step 2: If these LMIs have no solution, there exist two cases. One is that their solution is never found previously, and then, μ or/and α will be increased at a big step size. In general, μ is changed because the value range of μ ($\mu \geq 1$) is larger than that of α ($0 < \alpha < 1$). And go to Step 1. The other is that their solution is found last time, then, the previous μ or/and α are proper and go to step 4.
Step 3: If a solution is found, μ will be decreased at a small step size and go to step 1.
Step 4: μ and α are fixed, using similar idea to decrease γ such that γ is the optimized value. By (4.28), the parameters of a desired filter are given. EXIT.

Remark 4.11. It should be noted that due to the introduction of F_i^T and Ω_i, a desired filter can be easily constructed. It should also be pointed out that F_i^T is a diagonal matrix, which will be relaxed in the future research works. On the other hand, it may be possible to improve our criteria by adding more free variables based on the method in [74, 132]. As an expense, the computational load will be increased. Therefore, it is our future research topics to make a tradeoff between them. In addition, it is possible to extend our results to handle the nonlinear stochastic systems with time delays by means of the method in [94, 3].

4.2 Delay-Dependent Exponential H_∞ Model Reduction

In real engineering applications, many systems are generally characterized by high-order complex mathematical models, which bring serious difficulties to analysis and synthesis of the systems concerned. For example, to reduce the order of H_∞

filter is convenient to obtain the parameters of filter. Hence, model reduction issues have drawn considerable attention in recent decades. The objective of model reduction is to find a reduced-order model to approximate the original one such that the error between these two systems is small in some sense such as L_2, H_2 and H_∞. Many efficient approaches have been reported, for instance, the balanced truncation method [77], the optimal Hankel norm approximation method [31] and the aggregated method [60]. Very recently, LMIs technique is exploited to address the problem of H_∞ model reduction [143, 133, 130, 121, 29, 145]. In [145], the existence conditions for H_∞ model reduction for discrete-time switched systems are derived in terms of strict LMIs by using SLF method. However, time delays are not taken into account in [145]. When ADT method is used to address the problem of H_∞ model reduction for discrete-time switched systems with time delays, it requires the existence of some common Lyapunov matrices for all subsystems [144]. It is desired to construct a piecewise Lyapunov function, i.e. to select the corresponding Lyapunov matrices for each subsystem. To the best of the authors' knowledge, however, no result with regard to this problem is reported up to date. All the conditions obtained in [143, 133, 130, 121] are formulated in terms of LMIs with some non-convex constraints. These non-convex conditions make LMIs difficult to find the numerical solution, which is NP-hard[143]. Many optimization algorithms have been provided to solve such LMIs including the cone complementarity linearization (CCL) method [18] and sequential linear programming matrix method [61]. Nevertheless, these algorithms cause large amount of calculation and sometimes fail to guarantee the global convergence. Consequently, it is desired to derive the strict LMIs formulation. However, up to date, no result has been reported for model reduction issue of switched linear systems with time-varying delays under ADT switching signals, which is still open and remains unsolved. This motivates the present study.

In this section, the problem of exponential H_∞ model reduction for switched systems with a constant delay under ADT switching signals is investigated. Our objective is to overcome the restriction of some common Lyapunov matrices among all subsystems when ADT method is used to deal with the analysis of switched delay systems. Sufficient conditions for the existence of the desired reduced-order model are derived and formulated in terms of strict LMIs, which lead to lessen the computational complexity.

4.2.1 *Problem Formulation*

Consider discrete-time switched systems with state delays described by:

$$\begin{aligned} x_{k+1} &= A_\sigma x_k + A_{d\sigma} x_{k-d} + B_\sigma u_k, \\ y_k &= C_\sigma x_k + C_{d\sigma} x_{k-d} + D_\sigma u_k, \\ x_\psi &= \phi_\psi, \psi = -d, -d+1, \ldots, 1, \end{aligned} \tag{4.33}$$

where $x_k \in R^n$ is the state, $y_k \in R^m$ is the measured output, $u_k \in R^p$ is the input vector which belongs to $l_2[0,\infty)$. The piecewise constant function σ (denoting σ_k for simplicity) : $[0,\ \infty) \to \mathscr{P} = \{1, \cdots, p\}$ is a switching signal to specify, at each time instant k, the index $\sigma \in \mathscr{P}$ of the active subsystem, i.e., $\sigma = i$ means that the ith subsystem is activated. $p > 1$ is the number of subsystems. The ith subsystem is denoted by constant matrices A_i, A_{di}, B_i, C_i, C_{di} and D_i. $d > 0$ is a constant time delay.

In this section, we are interested in constructing a reduced-order switched system described by

$$\begin{aligned} \hat{x}_{k+1} &= A_{ri}\hat{x}_k + A_{rdi}\hat{x}_{k-d} + B_{ri}u_k, \\ \hat{y}_k &= C_{ri}\hat{x}_k + C_{rdi}\hat{x}_{k-d} + D_{ri}u_k, \end{aligned} \tag{4.34}$$

where $\hat{x}_k \in R^q$ is the state vector of the reduced-order system with $q < n$. $\hat{y}_k \in R^m$ is the output of reduced-order system with the same dimension m as the original output y_k. A_{ri}, B_{ri}, C_{ri} and D_{ri}, $i \in \mathscr{P}$ are the matrices with compatible dimensions to be determined. Note that system (4.34) is assumed to be switched synchronously by switching signal σ in system (4.33).

Remark 4.12. Note from (4.34) that the constant delay τ is assumed to be known a prior. Then, the conservatism can be reduced by extending the results to include the time-varying delays or the unknown delays. Another underlying assumption is to switch synchronously between the original systems and the reduced-order systems. A possible improvement to this point is to design an observer which allows the asynchronously switchings.

Augmenting the model of system (4.33) to include the states of (4.34), we can obtain the following model error system:

$$\begin{aligned} \tilde{x}_{k+1} &= \tilde{A}_i\tilde{x}_k + \tilde{A}_{di}\tilde{x}_{k-d} + \tilde{B}_iu_k, \\ \tilde{e}_k &= \tilde{C}_i\tilde{x}_k + \tilde{C}_{di}\tilde{x}_{k-d} + \tilde{D}_iu_k, \end{aligned} \tag{4.35}$$

where

$$\begin{aligned} &\tilde{x}_k = \begin{bmatrix} x_k \\ \hat{x}_k \end{bmatrix}, \ \tilde{e}_k = y_k - \hat{y}_k, \\ &\tilde{A}_i = \begin{bmatrix} A_i & 0 \\ 0 & A_{ri} \end{bmatrix}, \\ &\tilde{A}_{di} = \begin{bmatrix} A_{di} & 0 \\ 0 & A_{rdi} \end{bmatrix}, \ \tilde{B}_i = \begin{bmatrix} B_i \\ B_{ri} \end{bmatrix}, \\ &\tilde{C}_i = \begin{bmatrix} C_i & -C_{ri} \end{bmatrix}, \ \tilde{C}_{di} = \begin{bmatrix} C_{di} & -C_{rdi} \end{bmatrix}, \\ &\tilde{D}_i = D_i - D_{rdi}. \end{aligned}$$

Our objective is to design a reduced-order system in the form of system (4.34), and find admissible switching signals with ADT such that the resulting model error system (4.35) under such switching signals is exponentially stable and guarantees an H_∞ performance index.

To obtain our main result, the following lemma is presented.

Lemma 4.13. *([53]): For any constant matrix $M \geq 0$, $\Phi_l \in R^n$, there exist positive integers $\beta_1 \geq \beta_2 \geq 1$ such that*

$$-(\beta_2-\beta_1+1)\sum_{l=\beta_1}^{\beta_2}\Phi_l^T M\Phi_l \geq -\sum_{l=\beta_1}^{\beta_2}\Phi_l^T M\sum_{l=\beta_1}^{\beta_2}\Phi_l. \tag{4.36}$$

4.2.2 Design of Reduced-Order Model

In this section, the following theorem provides sufficient conditions for the existence of a reduced-order model (4.34) for system (4.33) based on Lemma 4.13.

Theorem 4.14. *For given scalars, $0<\alpha<1$, $\gamma>0$ and $\mu \geq 1$, system (4.35) is exponentially stable with an H_∞ norm bound γ under ADT switching signal σ, if there exist symmetric and positive definite matrices P_i, Q_i, Z_i and matrices M_{ci}, $c=1,2,3,\forall i\in\mathscr{P}$, such that the following inequalities hold,*

$$\begin{bmatrix}\Xi_1 & \Xi_2\\ * & \Xi_3\end{bmatrix}<0, \tag{4.37}$$

$$P_i \leq \mu P_j, Q_i \leq \mu Q_j, Z_i \leq \mu Z_j, \forall i,j\in\mathscr{P}, \tag{4.38}$$

and average dwell time τ_a

$$\tau_a>\tau^*=-\ln\mu/\ln\alpha, \tag{4.39}$$

where

$$\begin{aligned}
\Xi_1 &= \begin{bmatrix}\Psi_{11} & \Psi_{12} & \Psi_{13}\\ * & \Psi_{22} & \Psi_{23}\\ * & * & -\gamma^2 I\end{bmatrix},\\
\Xi_2 &= \begin{bmatrix}M_{1i} & \tilde{C}_i^T & \tilde{A}_i^T & A_i^T-I\\ M_{2i} & \tilde{C}_{di}^T & \tilde{A}_{di}^T & \tilde{A}_{di}^T\\ M_{3i} & \tilde{D}_i^T & \tilde{B}_i^T & \tilde{B}_i^T\end{bmatrix},\\
\Xi_3 &= diag\left\{-\alpha^d d^{-1}Z_i, -I, -P_i^{-1}, -(dZ_i)^{-1}\right\},\\
\Psi_{11} &= Q_i-\alpha P_i+M_{1i}^T+M_{1i},\\
\Psi_{12} &= M_{2i}^T-M_{1i},\\
\Psi_{22} &= -\alpha^{d_M}Q_i-M_{2i}^T-M_{2i},\\
\Psi_{23} &= M_{3i}^T,\ \Psi_{13}=M_{3i}^T.
\end{aligned}$$

Proof. Choosing the following Lyapunov-Krasovskii function as

$$V_i(k)=V_{1i}(k)+V_{2i}(k)+V_{3i}(k), \tag{4.40}$$

where

$$
\begin{aligned}
V_{1i}(k) &= \tilde{x}_k^T P_i \tilde{x}_k, \\
V_{2i}(k) &= \sum_{l=k-d}^{k-1} \alpha^{k-l-1} \tilde{x}_l^T Q_i \tilde{x}_l, \\
V_{3i}(k) &= \sum_{j=-d}^{-1} \sum_{l=k+j}^{k-1} \alpha^{k-l-1} \eta_l^T Z_{1i} \eta_l, \\
\eta_l &= \tilde{x}_{l+1} - \tilde{x}_l,
\end{aligned}
$$

and P_i, Q_i and Z_i are symmetric and positive definite matrices.

In terms of $\Delta V_i(k) = V_i(k+1) - \alpha V_i(k)$, we have

$$
\Delta V_{1i}(k) = \tilde{x}_{k+1}^T P_i \tilde{x}_{k+1} - \alpha \tilde{x}_k^T P_i \tilde{x}_k, \tag{4.41}
$$

$$
\Delta V_{2i}(k) = \tilde{x}_k^T Q_i \tilde{x}_k - \alpha^d \tilde{x}_{k-d}^T Q_i \tilde{x}_{k-d}, \tag{4.42}
$$

$$
\Delta V_{3i}(k) = d \eta_k^T Z_{1i} \eta_k - \sum_{l=k-d}^{k-1} \alpha^{k-l} \eta_l^T Z_{1i} \eta_l. \tag{4.43}
$$

By means of Lemma 4.13, we have

$$
\Delta V_{3i} \leq d \eta_k^T Z_{1i} \eta_k - \alpha^d d^{-1} \sum_{l=k-d}^{k-1} \eta_l^T Z_{1i} \sum_{l=k-d}^{k-1} \eta_l. \tag{4.44}
$$

Define $\xi_k = \begin{bmatrix} \tilde{x}_k^T & \tilde{x}_{k-d}^T & w_k^T \end{bmatrix}^T$, due to

$$
2\xi_k^T M_i \left[x_k - x_{k-d} - \sum_{l=k-d}^{k-1} \eta_l \right] = 0,
$$

one get

$$
\begin{aligned}
&\Delta V_i(k) + \tilde{e}_k^T \tilde{e}_k - \gamma^2 w_k^T w_k \\
&+ 2\xi_k^T M_i \left[x_k - x_{k-d_k} - \sum_{l=k-d_k}^{k-1} \eta_l \right] \\
&\leq \xi_k^T \left(\Xi_1 + M_i \alpha^{-d} d Z_i^{-1} M_i^T \right) \xi_k \\
&+ \tilde{z}_k^T \tilde{z}_k + \tilde{x}_{k+1}^T P_i \tilde{x}_{k+1} + \eta_k^T d Z_i \eta_k \\
&- \left[\xi_k^T M_i + \sum_{l=k-d}^{k-1} \alpha^d d^{-1} \eta_l^T Z_{1i} \right] (\alpha^d d^{-1} Z_{1i})^{-1} \\
&\times \left[M_i^T \xi_k + \sum_{l=k-d}^{k-1} \alpha^d d^{-1} Z_{1i} \eta_l \right].
\end{aligned} \tag{4.45}
$$

Due to $Z_i > 0$, the last term is non-positive. By Schur Complement, further, from (4.37), we have

$$\Delta V_i(k) + \tilde{e}_k^T \tilde{e}_k - \gamma^2 w_k^T w_k \leq 0. \tag{4.46}$$

Furthermore, one can obtain from (4.46) that

$$V_i(k+1) \leq \alpha V_i(k) - \Gamma(k). \tag{4.47}$$

where $\Gamma(k) = \tilde{e}_k^T \tilde{e}_k - \gamma^2 w_k^T w_k$.

The remain proof is similar to that of Theorem 4.3 and thus is omitted. □

Based on the above theorem, the parameters of the reduced-order systems are designed by the following theorem.

Theorem 4.15. *Consider system (4.33) and let* $0 < \alpha < 1$, $\gamma > 0$ *and* $\mu \geq 1$ *be given constants, if there exist symmetric and positive definite matrices* $P_i \in R^{m\times m}$, $Q_i \in R^{m\times m}$, $Z_i \in R^{m\times m}$, *and matrices* $X_i \in R^{n\times n}$, $Y_i \in R^{q\times q}$, $M_{si} \in R^{m\times m}$, $m = n+q, s = 1,2,3$, $\forall i \in \mathscr{P}$ *such that (4.38), ADT satisfies (4.39) and the following inequalities hold*

$$\begin{bmatrix} \Xi_1 & \tilde{\Xi}_4 \\ * & \tilde{\Xi}_5 \end{bmatrix} < 0, \tag{4.48}$$

where

$$\begin{aligned}
\tilde{\Xi}_4 &= \begin{bmatrix} M_{1i} & \tilde{\Psi}_{15}^T & \tilde{\Psi}_{16}^T & \tilde{\Psi}_{17}^T \\ M_{2i} & \tilde{\Psi}_{25}^T & \tilde{\Psi}_{26}^T & \tilde{\Psi}_{26}^T \\ M_{3i} & \tilde{\Psi}_{35}^T & \tilde{\Psi}_{36}^T & \tilde{\Psi}_{36}^T \end{bmatrix}, \\
\Omega_i &= \begin{bmatrix} X_i & 0 \\ Y_i^T E & Y_i \end{bmatrix}, \\
\tilde{\Xi}_5 &= diag\left\{ -\alpha^d d^{-1} Z_i, \ \ -I, \ \ \Psi_{66}, \ \ \Psi_{77} \right\}, \\
\tilde{\Psi}_{17} &= \tilde{\Psi}_{16} - \Omega_i, \\
\Psi_{66} &= P_i - \left(\Omega_i + \Omega_i^T\right), \\
\Psi_{77} &= dZ_i - \left(\Omega_i + \Omega_i^T\right), \\
\tilde{\Psi}_{15} &= \left[C_i \ -\bar{S}_i \right], \ \tilde{\Psi}_{25} = \left[C_{di} \ -\bar{N}_i \right], \\
\tilde{\Psi}_{35} &= D_i - \bar{R}_i, \ E = \left[I \ 0 \right], \\
\tilde{\Psi}_{16} &= \begin{bmatrix} X_i A_i & 0 \\ Y_i^T E A_i & L_i \end{bmatrix}, \\
\tilde{\Psi}_{26} &= \begin{bmatrix} X_i A_{di} & 0 \\ Y_i^T E A_{di} & H_i \end{bmatrix}, \\
\tilde{\Psi}_{36} &= \begin{bmatrix} X_i B_i \\ Y_i^T E B_i + T_i \end{bmatrix}.
\end{aligned}$$

Then, there exists an admissible reduced-order model (4.34) such that system (4.35) is exponentially stable with H_∞ norm bound γ under switching signals with ADT satisfying (4.39). In addition, if (4.38) and (4.48) have a feasible solution, switching signals with ADT satisfying (4.39) are found and the parameters of an admissible reduced-order model can be constructed by

$$\begin{bmatrix} A_{ri} & A_{rdi} & B_{ri} \\ C_{ri} & C_{rdi} & D_{ri} \end{bmatrix} = \begin{bmatrix} Y_i^{-1} & 0 \\ 0 & I \end{bmatrix} \begin{bmatrix} L_i & H_i & T_i \\ \bar{S}_i & \bar{N}_i & \bar{R}_i \end{bmatrix}. \tag{4.49}$$

Proof. By Theorem 4.14, system (4.35) under switching signals with ADT satisfying (4.39) is exponentially stable with H_∞ norm bound γ, if LMIs (4.37) and (4.38) hold.

For $P_i > 0$ and the nonsingular matrix Ω_i, from the fact that

$$(P_i - \Omega_i) P_i^{-1} (P_i - \Omega_i)^T \geq 0,$$

we have

$$-\Omega_i P_i^{-1} \Omega_i^T \leq P_i - \left(\Omega_i + \Omega_i^T\right).$$

Similarly, one has

$$-\Omega_i (dZ_i)^{-1} \Omega_i^T \leq dZ_i - \left(\Omega_i + \Omega_i^T\right).$$

Therefore, if the following inequality holds,

$$\begin{bmatrix} \Xi_1 & \Xi_4 \\ * & \tilde{\Xi}_5 \end{bmatrix} < 0, \tag{4.50}$$

where

$$\Xi_4 = \begin{bmatrix} M_{1i} & \tilde{C}_i^T & \tilde{A}_i^T \Omega_i^T & \Psi_{17} \\ M_{2i} & \tilde{C}_{di}^T & \tilde{A}_{di}^T \Omega_i^T & \tilde{A}_{di}^T \Omega_i^T \\ M_{3i} & \tilde{D}_i^T & \tilde{B}_i^T \Omega_i^T & \tilde{B}_i^T \Omega_i^T \end{bmatrix},$$
$$\Psi_{17} = (\tilde{A}_i^T - I) \Omega_i^T,$$

one can infer

$$\begin{bmatrix} \Xi_1 & \Xi_4 \\ * & \Xi_6 \end{bmatrix} < 0, \tag{4.51}$$

where

$$\Xi_6 = diag\left\{ -\alpha^d d^{-1} Z_i \;\; -I \;\; \hat{\Psi}_{66} \;\; \hat{\Psi}_{77} \right\},$$
$$\hat{\Psi}_{66} = -\Omega_i P_i^{-1} \Omega_i^T,$$
$$\hat{\Psi}_{77} = -\Omega_i (dZ_i)^{-1} \Omega_i^T.$$

Performing a congruence transformation to (4.51) via $diag\left\{I, I, I, I, I, \Omega_i^{-T}, \Omega_i^{-T}\right\}$ yields (4.37). In what follows, we will show (4.48) ensures that (4.50) holds.

Note that from (4.48), we have

$$\Omega_i + \Omega_i^T = \begin{bmatrix} X_i + X_i^T & E^T Y_i \\ * & Y_i + Y_i^T \end{bmatrix} > 0, \tag{4.52}$$

which means that X_i and Y_i are nonsingular. And thus, by using system (4.35), the following definitions are introduced,

$$\begin{aligned} &L_i \triangleq Y_i A_{ri},\ H_i \triangleq Y_i A_{rdi}, \\ &T_i \triangleq Y_i B_{ri},\ \bar{S}_i \triangleq C_{ri}, \\ &\bar{N}_i \triangleq C_{rdi}, \bar{R}_i \triangleq D_{ri}. \end{aligned} \tag{4.53}$$

By using (4.35) and (4.53), we can get

$$\begin{aligned} &\Omega_i \tilde{A}_i = \tilde{\Psi}_{110},\ \Omega_i \tilde{A}_{di} = \tilde{\Psi}_{210}, \\ &\Omega_i \tilde{B}_i = \tilde{\Psi}_{310},\ \tilde{C}_i = \tilde{\Psi}_{19},\ \tilde{C}_{di} = \tilde{\Psi}_{29}, \end{aligned}$$

and substitute them into (4.50) to have (4.48). This means that if (4.48) holds, (4.37) is true, which implies that the error system (4.35) is exponentially stable with an H_∞ performance index. Meanwhile, from (4.53) that the parameters of a reduced-order model are given by (4.49). The proof is completed. □

Remark 4.16. It should be pointed out that the Lyapunov function (4.40) is general with comparison to the existing Lyapunov function. When $\alpha = 1$ in (4.40), the Lyapunov function (4.40) reduces to those in [29] for non-switched delay systems. When $\alpha = 0$, the Lyapunov function (4.40) is capable of dealing with switched systems [144]. It is obvious to see from Ξ_2 and Ξ_3 in (4.37) that it is difficult to deal with the problem of model reduction due to the existence of P_i^{-1}, which leads to some product terms between P_i and $\tilde{A}_i$, $\tilde{A}_{di}$ and $\tilde{B}_i$. There exists the same case between Z_i and $\tilde{A}_i$, $\tilde{A}_{di}$ and $\tilde{B}_i$. To overcome the difficulties, motivated by the work [82], an auxiliary slack matrix Ω_i is introduced in the proof of Theorem 1 such that these product terms are decoupled. That is, P_i and Z_i are not involved in any product with $\tilde{A}_i$, $\tilde{A}_{di}$ and $\tilde{B}_i$ in (4.50). This makes it feasible to construct a reduced-order model. On the other hand, the introduction of the auxiliary slack matrices leads to the use of so much resources such as computer storage space. Therefore, it is the future topic to make the proper tradeoff between them.

Remark 4.17. Note that many existing results on model reduction in the discrete-time context are formulated in terms of LMIs with inverse constraints or other non-convex conditions, (see, for instance, [29, 130] and references therein), which is difficult to obtain the numerical solutions. An LMI-based solution is obtained for the model reduction problem of switched systems in [145], but the time delay is not taken into account. The main reason is that the existence of time delay will lead to the common Lyapunov matrices, which is the conservatism [144]. To reduce the conservatism, a factor α is introduced such that all Lyapunov matrices are piecewise.

4.3 Illustrative Examples

In this section, two numerical examples are provided to illustrate the effectiveness of the proposed method. One is for filter design and the other is for reduced-order model design.

Example 4.18. Consider discrete-time switched system (4.1) consisting of two subsystems, with subsystem 1:

$$\begin{aligned}
A_1 &= \begin{bmatrix} 0.2 & -0.1 \\ 0.1 & 0.4 \end{bmatrix}, A_{d1} = \begin{bmatrix} 0.1 & 0 \\ 0.1 & 0.1 \end{bmatrix}, \\
B_1^T &= \begin{bmatrix} 0 & 0.1 \end{bmatrix}, C_1 = \begin{bmatrix} 0.1 & 0 \end{bmatrix}, \\
C_{d1} &= \begin{bmatrix} 0 & 0.1 \end{bmatrix}, D_1 = 0.1, \\
G_1 &= \begin{bmatrix} 0.4 & 0.6 \end{bmatrix}, G_{d1} = \begin{bmatrix} 0.1 & 0 \end{bmatrix}, \\
E_1 &= 0.01.
\end{aligned}$$

and subsystem 2:

$$\begin{aligned}
A_2 &= \begin{bmatrix} 0.4 & 0.2 \\ -0.1 & 0.3 \end{bmatrix}, A_{d2} = \begin{bmatrix} 0.1 & 0 \\ -0.01 & 0.1 \end{bmatrix}, \\
B_2^T &= \begin{bmatrix} 0 & 0.2 \end{bmatrix}, C_2 = \begin{bmatrix} 0.2 & 0 \end{bmatrix}, \\
C_{d1} &= \begin{bmatrix} 0 & 0.1 \end{bmatrix}, D_2 = 0.2, \\
G_2 &= \begin{bmatrix} 0.3 & 0.2 \end{bmatrix}, G_{d2} = \begin{bmatrix} 0 & 0.1 \end{bmatrix}, \\
E_2 &= 0.02.
\end{aligned}$$

Let $d_m = 1$, $d_M = 4$, $\alpha = 0.6$, $\mu = 1.05$ and $\gamma = 0.8$, by utilizing the LMI Toolbox, it follows from Theorem 4.5 that we can have a feasible solution. Furthermore, by (4.28), the desired filter can be obtained with the parameterized matrices as follows:

$$\left[\begin{array}{c|c} A_{f1} & B_{f1} \\ \hline C_{f1} & D_{f1} \end{array}\right] = \left[\begin{array}{cc|c} 0.1346 & -0.1724 & -2.3715 \\ 0.0124 & 0.2667 & -9.0656 \\ \hline 0.0181 & 0.0613 & 0.8871 \end{array}\right],$$

$$\left[\begin{array}{c|c} A_{f2} & B_{f2} \\ \hline C_{f2} & D_{f2} \end{array}\right] = \left[\begin{array}{cc|c} 0.4079 & 0.2898 & -0.1514 \\ -0.1634 & 0.1926 & -5.5188 \\ \hline 0.0333 & 0.0356 & 0.2484 \end{array}\right],$$

and $\tau^* = 0.0955$ is given by (4.7). The different α and τ^* for $\mu = 1.01$ and $\gamma = 0.3$ are shown in Table 4.1. As $\alpha \to 1$, $\tau^* \to \infty$, which means that switched delay systems will be changed to classic delay systems. While $\alpha \to 0$, no solution is found because it is suitable for switched systems instead of switched delay systems. To illustrate the performances of the designed filter, we choose $w_k = 5e^{-0.1k}cos(0.2\pi k)$, which belongs to $l_2[0,\infty)$. Under the zero-initial condition, given $\mu = 1.05, \gamma = 0.8, \alpha = 0.6$, the simulation results are shown in Fig. 4.1-Fig. 4.4, with ADT switching signal satisfying $\tau_a = 0.9272$. From the simulation curves above, it can be clearly observed that the designed filter can accomplish the filtering task.

Example 4.19. Consider discrete-time switched system (4.33) consisting of two subsystems, with parameters from [29] and [145]:

$$A_1 = \rho \begin{bmatrix} 1.3 & 2.2 & -1.3 & 0.8 \\ 0.5 & -0.3 & 1.9 & 0.6 \\ -0.7 & -0.5 & -0.4 & -1.2 \\ -1.7 & 2.1 & 0.3 & 2.8 \end{bmatrix},$$

$$A_2 = \rho \begin{bmatrix} 1.1 & 2.2 & -1.3 & 0.8 \\ 0.5 & -0.3 & 1.5 & 0.6 \\ -0.7 & -0.3 & -0.4 & -1.2 \\ -1.7 & 2.1 & 0.3 & 2.0 \end{bmatrix},$$

$$A_{d1} = A_{d2} = \rho \begin{bmatrix} 0.2 & 0.1 & 0 & 0 \\ 0 & 0.2 & 0 & 0 \\ 0 & 0 & 0.2 & 0.1 \\ 0 & 0 & 0 & 0.2 \end{bmatrix},$$

$$B_1 = \rho \begin{bmatrix} 1.9 \\ -1.8 \\ 1.6 \\ -0.8 \end{bmatrix}, \; B_2 = \rho \begin{bmatrix} 2.3 \\ -1.3 \\ 1.6 \\ -0.4 \end{bmatrix},$$

$$C_1 = \rho \begin{bmatrix} 12 & 5 & 0.3 & 2.8 \end{bmatrix},$$
$$C_2 = \rho \begin{bmatrix} 12 & 5 & 0.3 & 2.8 \end{bmatrix},$$
$$C_{d1} = C_{d2} = \rho \begin{bmatrix} 0.2 & 0.5 & 0.1 & 0.9 \end{bmatrix},$$
$$D_1 = D_2 = \rho, \; d = 2.$$

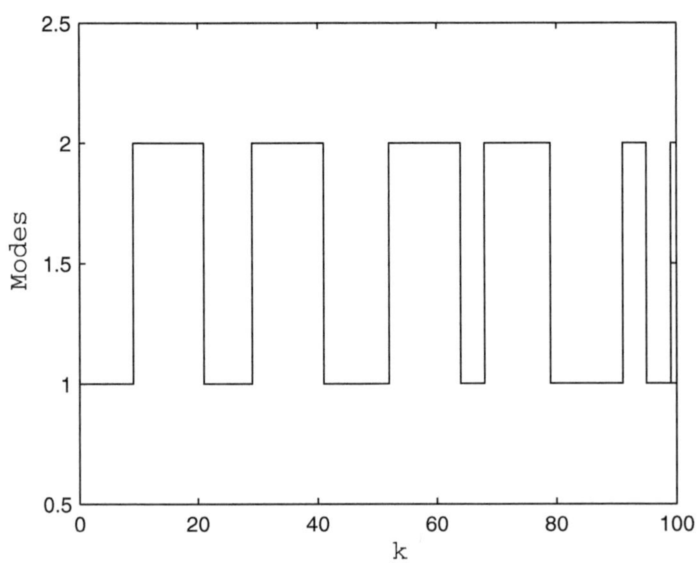

Fig. 4.1 Switching signal

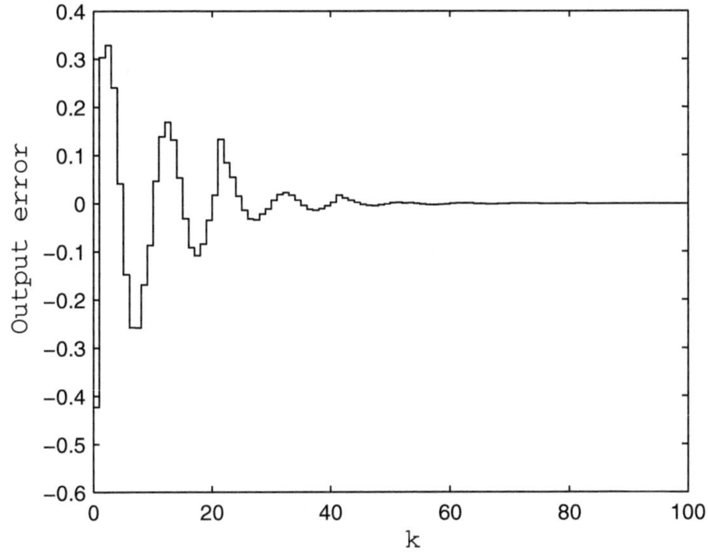

Fig. 4.2 The output error $z_k - \hat{z}_k$

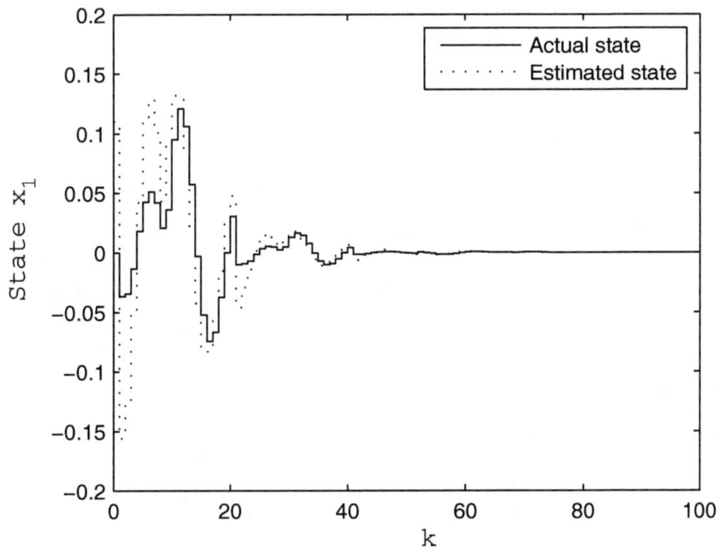

Fig. 4.3 The state x_1 and its estimation

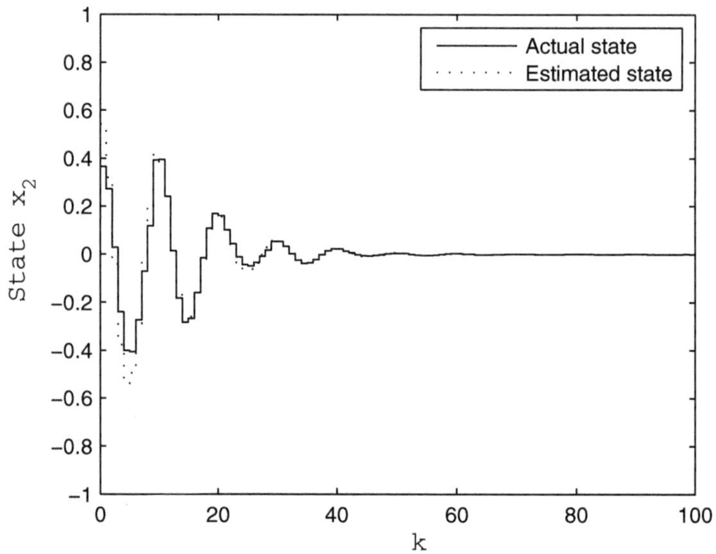

Fig. 4.4 The state x_2 and its estimation

Table 4.1 The different α and τ^* for $\mu = 1.01$ and $\gamma = 0.3$

α	0.3	0.5	0.9	0.9999
τ^*	–	0.0144	0.0944	99.4983

The parameter A_{di} of time delay term stems from those of [29] while others are from those in [145] since the latter is concerned with switched delay-free systems. Here, we are interested in designing a q-order ($q < 4$) system (4.34) and find out an admissible ADT switching signals such that the model error system (4.35) is exponentially stable with H_∞ norm bound γ. Given $\rho = 0.1$, $\alpha = 0.9$, $\mu = 1.2$ and $\gamma = 2.0$, by utilizing the LMI Toolbox, it follows from Theorem 4.15 that the following three reduced-order models can be given.

The Third-Order Model Is :

$$A_{r1} = \begin{bmatrix} 0.2610 & 0.0527 & -0.0216 \\ 0.0475 & 0.2262 & 0.0234 \\ -0.0099 & -0.0128 & 0.2653 \end{bmatrix},$$

$$A_{dr1} = \begin{bmatrix} 0.0008 & 0.0026 & -0.0021 \\ 0.0059 & -0.0013 & 0.0037 \\ 0.0008 & -0.0007 & 0.0047 \end{bmatrix},$$

$$B_{r1} = \begin{bmatrix} -0.3226 & 0.5332 & -0.3734 \end{bmatrix}^T,$$
$$C_{r1} = \begin{bmatrix} -0.0618 & 0.0461 & -0.0118 \end{bmatrix},$$
$$C_{dr1} = \begin{bmatrix} -0.0125 & 0.0126 & -0.0113 \end{bmatrix},$$
$$D_{r1} = -0.7156,$$

and

$$A_{r2} = \begin{bmatrix} 0.2554 & 0.0497 & -0.0243 \\ 0.0322 & 0.2466 & 0.0058 \\ -0.0117 & -0.0147 & 0.2678 \end{bmatrix},$$
$$A_{dr2} = \begin{bmatrix} 0.0006 & 0.0028 & -0.0021 \\ 0.0058 & -0.0012 & 0.0039 \\ -0.0011 & 0.0014 & 0.0027 \end{bmatrix},$$
$$B_{r2} = \begin{bmatrix} -0.6795 & 0.1368 & -0.5570 \end{bmatrix}^T,$$
$$C_{r2} = \begin{bmatrix} -0.0452 & 0.0055 & -0.0121 \end{bmatrix},$$
$$C_{dr2} = \begin{bmatrix} -0.0039 & 0.0071 & -0.0053 \end{bmatrix},$$
$$D_{r2} = -0.3804.$$

The Second-Order Model Is :

$$\left[\begin{array}{c|c|c} A_{r1} & A_{rd1} & B_{r1} \\ \hline C_{r1} & C_{rd1} & D_{r1} \end{array}\right] = \left[\begin{array}{cc|cc|c} 0.2719 & 0.0399 & 0.0000 & 0.0032 & -0.5025 \\ 0.0292 & 0.2509 & 0.0043 & 0.0005 & 0.6031 \\ \hline -0.0661 & 0.0372 & -0.0083 & 0.0112 & -0.6833 \end{array}\right],$$

and

$$\left[\begin{array}{c|c|c} A_{r2} & A_{rd2} & B_{r2} \\ \hline C_{r2} & C_{rd2} & D_{r2} \end{array}\right] = \left[\begin{array}{cc|cc|c} 0.2711 & 0.0380 & 0.0007 & 0.0027 & -0.7842 \\ 0.0242 & 0.2563 & 0.0035 & 0.0009 & 0.4300 \\ \hline -0.0600 & -0.0140 & -0.0004 & 0.0056 & -0.2979 \end{array}\right].$$

The First-Order Model Is:

$$\left[\begin{array}{c|c|c} A_{r1} & A_{rd1} & B_{r1} \\ \hline C_{r1} & C_{rd1} & D_{r1} \end{array}\right] = \left[\begin{array}{c|c|c} 0.2532 & -0.0024 & -0.7170 \\ \hline -0.0600 & -0.0086 & -0.6759 \end{array}\right],$$

and

$$\left[\begin{array}{c|c|c} A_{r2} & A_{rd2} & B_{r2} \\ \hline C_{r2} & C_{rd2} & D_{r2} \end{array}\right] = \left[\begin{array}{c|c|c} 0.2502 & -0.0018 & -1.0418 \\ \hline -0.0496 & 0.0006 & -0.3038 \end{array}\right].$$

In addition, the minimal ADT $\tau^* = 1.7305$ is obtained by (4.39).

4.4 Summary

We have studied the problem of exponential H_∞ filter design for discrete-time switched delay systems with interval time-varying delay under ADT switching signals. By introducing a factor α combining with ADT method, sufficient conditions for the solvability of this problem have been established in terms of LMIs. A desired filter has been designed by introducing two kinds of auxiliary slack matrix variables. Since some practical systems are high-order, we have studied the problem of the exponential H_∞ model reduction for discrete-time switched systems under ADT switching signals as well. Based on H_∞ performance analysis and new linearization technique, sufficient conditions for the solvability of this problem have been established in terms of strict LMIs, which avoid the numerous work of the calculation and decrease the computation complexity. Two examples have been given to show the effectiveness of the proposed methods.

It is an interesting research topic to extend the proposed approach to controller design for the systems concerned. Due to the introduction of α, ADT method is able to be applicable to analyze and synthesize for discrete time switched delay systems under switching law with ADT such as H_2/H_∞ control and L_2/L_∞ control.

5

Exponential H_∞ Criteria for Continuous-Time Switched Delay Systems

This chapter deals with the problem of exponential H_∞ filtering for a class of continuous-time switched linear system with interval time-varying delay. The time delay under consideration includes two cases: one is that the time delay is differentiable and bounded with a constant delay-derivative bound while the other is that the time delay is continuous and bounded. Switched linear filters are designed to ensure that the filtering error systems under switching signal with average dwell time are exponentially stable with a prescribed H_∞ noise attenuation level. Based on the results of filter design, the problem of model reductions for switched systems with two different kinds of time delays is also studied and sufficient conditions for the existence of the reduced-order model are derived in terms of LMIs. Finally, four numerical examples are presented to illustrate the effectiveness of the proposed techniques.

5.1 Delay-Dependent Exponential H_∞ Filtering

In the previous chapters, H_∞ filtering problems for switched delay systems are studied in the discrete-time context. An exponential H_∞ filter for a class of polytopic uncertain discrete-time switched linear systems with average dwell time switching is designed by means of a Lyapunov function approach in [141]. In real physical devices or plants, the system states vary continuously. If the systems are not sampled or discretized, their dynamics are continuous. Up to date, to the best of the authors' knowledge, the exponential H_∞ filtering problem has not been addressed for continuous-time switched systems with an interval time-varying delay. In the previous chapter, a weighting factor is introduced such that ADT method can be used to obtain a kind of slow switching signal.

In this chapter, the problem of exponential H_∞ filtering for continuous-time switched systems with an interval time-varying delay is investigated. First, the interval time-varying delay with a constant delay-derivative bound is considered. By means of the free weighting matrix approach and the average dwell time technology, a filter is developed to ensure that the filtering error systems under switching signal with average dwell time are exponentially stable with a prescribed H_∞ noise attenuation level. Secondly, the obtained results above are extended to the case that

D. Wang et al.: *Robust Filtering & Fault Detect. of Switched Delay Systems*, LNCIS 445, pp. 75–103.
DOI: 10.1007/978-3-642-37685-6_5

the interval time-varying is in absence of the constant delay-derivative bound. In the end, two illustrative examples are presented to demonstrate the feasibility and efficiency of the proposed filter.

The rest of this chapter is organized as follows. In Section 5.1.1, the problem under consideration is formulated. The H_∞ filtering performance is analyzed in Section 5.1.2. Sufficient conditions on the existence of the H_∞ filters for discrete-time switched systems with state delays are derived in terms of LMI and the desired filters are constructed in Section 5.1.3. To demonstrate the validity of the proposed approach, an example is given in Section 5.1.4 which is followed by a conclusion in Section 5.1.5.

5.1.1 Problem Formulation

Consider a class of switched linear systems with state delay:

$$\begin{aligned} \dot{x}(t) &= A_\sigma x(t) + E_\sigma x(t-d(t)) + B_\sigma w(t), \\ y(t) &= C_\sigma x(t) + F_\sigma x(t-d(t)) + D_\sigma w(t), \\ z(t) &= H_\sigma x(t) + G_\sigma x(t-d(t)) + M_\sigma w(t), \\ x(\theta) &= \phi(\theta), \quad \theta \in [-h_2, 0], \end{aligned} \tag{5.1}$$

where $x(t) \in R^n$ is the state, $y(t) \in R^r$ is the measured output, $z(t) \in R^s$ is the signal to be estimated, $w(t) \in R^q$ is the disturbance input which is assumed to belong to $l_2[0,\infty)$. The piecewise constant function σ denoting $\sigma(t)$ for simplicity : $[0, \infty) \to \mathscr{P} = \{1, \cdots, p\}$ is a switching signal to specify, at each time instant t, the index $\sigma \in \mathscr{P}$ of the active subsystem, i.e., $\sigma = i$ means that the ith subsystem is activated. $p > 1$ is the number of subsystems. The ith subsystem is denoted by constant matrices $A_i, E_i, B_i, C_i, F_i, D_i, H_i, G_i, M_i$ with the appropriate dimensions. $d(t)$ is assumed to be the interval time-varying delay satisfying the following two cases, respectively,

(A1): $d(t)$ is differentiable and bounded with a constant delay-derivative bound,

$$h_1 \le d(t) \le h_2, \ \dot{d}(t) \le d, \tag{5.2}$$

(A2): $d(t)$ is continuous and bounded,

$$h_1 \le d(t) \le h_2, \tag{5.3}$$

where $h_2 > h_1 \ge 0$ are scalars representing the upper and the lower bounds of the time delay, respectively.

Remark 5.1. It is clear that (A2) contains (A1) as a special case. It is only required that the time-varying delay is a continuous and bounded function in (A2) while it is assumed that the time-varying delay is differentiable and its derivative is bounded in (A1). Thus, the result obtained under (A1) is less conservative than that under (A2), if (A1) is satisfied. However, if the time-varying delay is not differentiable, only (A2) can be used to deal with the situation.

In this chapter, a filter we shall design is described by

$$\begin{aligned}\dot{\hat{x}}(t) &= A_{fi}\hat{x}(t) + B_{fi}y(t),\\ \hat{z}(t) &= C_{fi}\hat{x}(t) + D_{fi}y(t),\end{aligned} \tag{5.4}$$

where $\hat{x}(t) \in R^n$ and $\hat{z}(t) \in R^s$ and $A_{fi}, B_{fi}, C_{fi}, D_{fi}$, $\forall i \in \mathscr{P}$ are the parameterized filer matrices to be determined. It is assumed that the filter (5.4) is switched synchronously by the switching signal σ in system (5.1). This means that the switched signal in filter (5.4) is the same as that in system (5.1). Thus, the filter in the form of (5.4) is itself a switched system.

By introducing $\tilde{x}(t) = \begin{bmatrix} x^T(t) & \hat{x}^T(t) \end{bmatrix}^T$ and $\tilde{z}(t) = z(t) - \hat{z}(t)$, and augmenting system (5.1) to include switched linear filter (5.4), we can obtain the filtering error system

$$\begin{aligned}\dot{\tilde{x}}(t) &= \tilde{A}_\sigma \tilde{x}(t) + \tilde{E}_\sigma K\tilde{x}(t-d(t)) + \tilde{B}_\sigma w(t),\\ \tilde{z}(t) &= \tilde{H}_\sigma \tilde{x}(t) + \tilde{G}_\sigma K\tilde{x}(t-d(t)) + \tilde{M}_\sigma w(t),\end{aligned} \tag{5.5}$$

where

$$\begin{aligned}\tilde{A}_\sigma &= \begin{bmatrix} A_\sigma & 0 \\ B_{f\sigma}C_\sigma & A_{f\sigma} \end{bmatrix},\\ \tilde{E}_\sigma &= \begin{bmatrix} E_\sigma \\ B_{f\sigma}F_\sigma \end{bmatrix},\\ \tilde{B}_\sigma &= \begin{bmatrix} B_\sigma \\ B_{f\sigma}D_\sigma \end{bmatrix},\\ \tilde{H} &= \begin{bmatrix} H_\sigma - D_{f\sigma}C_\sigma & -C_{f\sigma} \end{bmatrix},\\ \tilde{G}_\sigma &= G_\sigma - D_{f\sigma}F_\sigma,\\ \tilde{M}_\sigma &= M_\sigma - D_{f\sigma}D_\sigma,\ K = \begin{bmatrix} I & 0 \end{bmatrix}.\end{aligned}$$

To present our objective more precisely, the following definitions are introduced.

Definition 5.2. *The equilibrium point $x^* = 0$ of system (5.1) is said to be exponentially stable under the switching signal σ if the solution of system (5.1) satisfies*

$$\|x(t)\| \le \gamma \|x(t_0)\|_\theta e^{-\lambda(t-t_0)}, \quad \forall t \ge t_0, \tag{5.6}$$

for all $\gamma \ge 1$ and $\lambda > 0$, where $\|\bullet\|$ denotes the Euclidean norm and $\|x(t)\|_\theta$ is defined by $\|x(t)\|_\theta = \sup\limits_{-h_2 \le \theta \le 0} \{\|x(t+\theta)\|, \|\dot{x}(t+\theta)\|\}$.

Definition 5.3. *Constructing a switching sequence*

$$\{(\sigma(t_0), t_0), \cdots, (\sigma(t_k), t_k) \,|k = 1, 2, \cdots\},$$

where $(\sigma(t_k), t_k)$ means the $\sigma(t_k)$-th subsystem is activated during $t \in [t_k, t_{k+1})$. k denotes the switching ordinal number. For given scalars $\tau^* > 0$ and $N_0 \ge 0$, we have

$$N_\sigma(t_0,t) \le N_0 + \frac{t-t_0}{\tau^*}, \tag{5.7}$$

where $N_\sigma(t_0,t)$ denotes the switching number of σ during (t_0,t). Then, τ^* and N_0 are called average dwell time and the chattering bound, respectively. Here we assume $N_0 = 0$ for simplicity as commonly used in literature.

Remark 5.4. Switching signals with average dwell time means that the time interval of switching signals between consecutive switching is at least τ^* on the average. This is a kind of slowly switching signals, which is less conservative than arbitrary switching signals.

The exponential H_∞ filtering problem to be addressed is formulated as follows. Given switched linear system (5.1) and a prescribed level of noise attenuation $\gamma > 0$, determine a filter in the form of (5.4) and find out admissible switching signals with average dwell time such that the following requirements are satisfied:

(1) the filtering error system (5.5) with (A1) or (A2) is exponentially stable when $w(t) = 0$;
(2) the H_∞ performance $\|\tilde{z}(t)\|_2 \le \gamma \|w(t)\|_2$, under zero-initial condition, is guaranteed for all nonzero $w(t) \in l_2[0,\infty)$.

In this situation, the filtering error system (5.5) with (A1) or (A2) is said to be exponentially stable with an H_∞ prescribed attenuation level γ.

5.1.2 Stability and H_∞ Performance Analysis

In this section, delay-dependent sufficient conditions on the existence of a filter, which guarantees that the filtering error system (5.5) with (A1) or (A2) is exponentially stable with an H_∞ prescribed attenuation level γ, are given. We first focus our attention on the case (A1) because the results under (5.1) are easily extended to those under (A2). Secondly, the case (A2) is studied. For simplicity, the subscript of the *ith* subsystem of the filtering error system (5.5) under (A1) is omitted and rewritten as follows:

$$\begin{aligned}\dot{\tilde{x}}(t) &= \tilde{A}\tilde{x}(t) + \tilde{E}K\tilde{x}(t-d(t)) + \tilde{B}w(t),\\ \tilde{z}(t) &= \tilde{H}\tilde{x}(t) + \tilde{G}K\tilde{x}(t-d(t)) + \tilde{M}w(t),\end{aligned} \tag{5.8}$$

and construct the corresponding Lyapunov function candidate as

$$\begin{aligned} V(t) = {} & \tilde{x}^T(t)P\tilde{x}(t) + e^{-\alpha t}\sum_{l=1}^{2}\int_{t-h_l}^{t}\tilde{x}^T(s)e^{\alpha s}K^T Q_l K\tilde{x}(s)\,ds \\ & + e^{-\alpha t}\int_{t-d(t)}^{t}\tilde{x}^T(s)e^{\alpha s}K^T Q_3 K\tilde{x}(s)\,ds \\ & + e^{-\alpha t}\int_{-h_2}^{0}\int_{t+\theta}^{t}\dot{\tilde{x}}^T(s)e^{\alpha s}K^T Z_1 K\dot{\tilde{x}}(s)\,ds\,d\theta \\ & + e^{-\alpha t}\int_{-h_2}^{-h_1}\int_{t+\theta}^{t}\dot{\tilde{x}}^T(s)e^{\alpha s}K^T Z_2 K\dot{\tilde{x}}(s)\,ds\,d\theta, \end{aligned} \tag{5.9}$$

where $P > 0$, $Z_l > 0$, $Q_k > 0, k = 1,2,3, l = 1,2$ are symmetric matrices. Thus, we will get the following lemma which plays a key role in our derivation.

Lemma 5.5. *Consider system (5.8) under (A1), let $\alpha > 0$, $d > 0$ and $h_2 > h_1 \geq 0$ be given scalars, if there exist symmetric and positive definite matrices P_l, Z_l, Q_k, and matrices $Y_k, T_k, L_k, k = 1,2,3, l = 1,2$ such that the following inequalities hold*

$$P_1 - P_2 > 0, \tag{5.10}$$

$$\begin{bmatrix} \Omega & R \\ * & -W \end{bmatrix} < 0, \tag{5.11}$$

then, we have

$$V(t) \leq e^{-\alpha(t-t_0)} V(t_0) - \int_{t_0}^{t} e^{-\alpha(t-s)} \Gamma(s)\, ds, \tag{5.12}$$

where

$$\Omega = \begin{bmatrix} \varphi_{11} & \varphi_{12} & \varphi_{13} & \varphi_{14} \\ * & \varphi_{22} & \varphi_{23} & \varphi_{24} \\ * & * & \varphi_{33} & \varphi_{34} \\ * & * & * & -\gamma^2 I \end{bmatrix},$$

$$R = \begin{bmatrix} L_1 & -T_1 & A^T U & H^T - C^T D_f^T & Y_1 & T_1 & L_1 \\ 0 & 0 & 0 & -C_f^T & 0 & 0 & 0 \\ L_2 & -T_2 & E^T U & \tilde{G}^T & Y_2 & T_2 & L_2 \\ L_3 & -T_3 & B^T U & \tilde{M}^T & Y_3 & T_3 & L_3 \end{bmatrix},$$

$$W = diag\left\{ e^{-\alpha h_1} Q_1,\ e^{-\alpha h_2} Q_2,\ U,\ I,\ H_{20}^{-1} Z_1,\ H_{21}^{-1} Z_3,\ H_{21}^{-1} Z_2 \right\},$$

$$H_{21} = \frac{e^{\alpha h_2} - e^{\alpha h_1}}{\alpha},\ H_{20} = \frac{e^{\alpha h_2} - 1}{\alpha},\ h_{21} = h_2 - h_1,$$

$$U = (h_2 Z_1 + h_{21} Z_2),\ Z_3 = Z_1 + Z_2,$$

$$\varphi_{11} = P_1 A - P_2 B_f C + A^T P_1 - C^T B_f^T P_2 + \sum_{i=1}^{3} Q_i + Y_1 + Y_1^T + \alpha P_1,$$

$$\varphi_{12} = -P_2 A_f - A^T P_2 + C^T B_f^T P_2 - \alpha P_2,$$

$$\varphi_{22} = P_2 A_f + A_f^T P_2 + \alpha P_2,$$

$$\varphi_{13} = P_1 E - P_2 B_f F - Y_1 + Y_2^T + T_1 - L_1,$$

$$\varphi_{23} = -P_2 E + P_2 B_f F,$$

$$\varphi_{33} = (d-1) e^{-\alpha h_1} Q_3 - Y_2 - Y_2^T + T_2 + T_2^T - L_2 - L_2^T,$$

$$\varphi_{14} = P_1 B - P_2 B_f D + Y_3^T,$$

$$\varphi_{24} = -P_2 B + P_2 B_f D,$$

$$\varphi_{34} = T_3^T - Y_3^T - L_3^T,$$

$$\Gamma(s) = \tilde{z}^T(s) \tilde{z}(s) - \gamma^2 w^T(s) w(s).$$

Proof. Taking the derivative of (5.9), we get

$$\begin{aligned}
\dot{V}(t) &= 2\tilde{x}^T(t)P\dot{\tilde{x}}(t)+\tilde{x}^T(t)K^TQ_1K\tilde{x}(t)\\
&-\tilde{x}^T(t-h_1)e^{-\alpha h_1}K^TQ_1K\tilde{x}(t-h_1)+\tilde{x}^T(t)K^TQ_2K\tilde{x}(t)\\
&-\tilde{x}^T(t-h_2)e^{-\alpha h_2}K^TQ_2K\tilde{x}(t-h_2)+\tilde{x}^T(t)K^TQ_3K\tilde{x}(t)\\
&-\left(1-\dot{d}(t)\right)\tilde{x}^T(t-d(t))e^{-\alpha d(t)}K^TQ_3K\tilde{x}(t-d(t))\\
&-e^{-\alpha t}\int_{t-h_2}^{t}\dot{\tilde{x}}^T(s)e^{\alpha s}K^TZ_1K\dot{\tilde{x}}(s)ds+h_2\dot{\tilde{x}}^T(t)K^TZ_1K\dot{\tilde{x}}(t)\\
&+h_2\dot{\tilde{x}}^T(t)K^TZ_2K\dot{\tilde{x}}(t)-h_1\dot{\tilde{x}}^T(t)K^TZ_2K\dot{\tilde{x}}(t)+\alpha\tilde{x}^T(t)P\tilde{x}(t)\\
&-e^{-\alpha t}\int_{t-h_2}^{t-h_1}\dot{\tilde{x}}^T(s)e^{\alpha s}K^TZ_2K\dot{\tilde{x}}(s)ds-\alpha V(t).
\end{aligned}\tag{5.13}$$

From the Leibniz-Newton formula, the following equations are true:

$$\begin{aligned}
&2\left[\tilde{x}^T(t)K^TY_1+\tilde{x}^T(t-d(t))K^TY_2+w^T(t)Y_3\right]\\
&\times\left[K\tilde{x}(t)-K\tilde{x}(t-d(t))-\int_{t-d(t)}^{t}K\dot{\tilde{x}}(s)ds\right]=0,
\end{aligned}\tag{5.14}$$

$$\begin{aligned}
&2\left[\tilde{x}^T(t)K^TT_1+\tilde{x}^T(t-d(t))K^TT_2+w^T(t)T_3\right]\\
&\times\left[K\tilde{x}(t-d(t))-K\tilde{x}(t-h_2)-\int_{t-h_2}^{t-d(t)}K\dot{\tilde{x}}(s)ds\right]=0,
\end{aligned}\tag{5.15}$$

$$\begin{aligned}
&2\left[\tilde{x}^T(t)K^TL_1+\tilde{x}^T(t-d(t))K^TL_2+w^T(t)L_3\right]\\
&\times\left[K\tilde{x}(t-h_1)-K\tilde{x}(t-d(t))-\int_{t-d(t)}^{t-h_1}K\dot{\tilde{x}}(s)ds\right]=0,
\end{aligned}\tag{5.16}$$

where $Y_k, T_k, L_k, k=1,2,3$ are the free weighting matrices with appropriate dimensions. They are used to express the relationship between the terms in the Leibniz-Newton formula and can be determined by solving the corresponding LMIs.

For the integration terms in (5.13), we can break them up as follows:

$$\begin{aligned}
&e^{-\alpha t}\int_{t-h_2}^{t}\dot{\tilde{x}}^T(s)e^{\alpha s}K^TZ_1K\dot{\tilde{x}}(s)ds\\
&=\int_{t-d(t)}^{t}\dot{\tilde{x}}^T(s)e^{\alpha(s-t)}K^TZ_1K\dot{\tilde{x}}(s)ds\\
&+\int_{t-h_2}^{t-d(t)}\dot{\tilde{x}}^T(s)e^{\alpha(s-t)}K^TZ_1K\dot{\tilde{x}}(s)ds,
\end{aligned}\tag{5.17}$$

and

$$
\begin{aligned}
& e^{-\alpha t}\int_{t-h_2}^{t-h_1} \dot{\tilde{x}}^T(s)\, e^{\alpha s} K^T Z_2 K \dot{\tilde{x}}(s)\, ds \\
&= \int_{t-h_2}^{t-d(t)} \dot{\tilde{x}}^T(s)\, e^{\alpha(s-t)} K^T Z_2 K \dot{\tilde{x}}(s)\, ds \\
&+ \int_{t-d(t)}^{t-h_1} \dot{\tilde{x}}^T(s)\, e^{\alpha(s-t)} K^T Z_2 K \dot{\tilde{x}}(s)\, ds. \qquad (5.18)
\end{aligned}
$$

Thus, from (5.14) and (5.17), we can obtain that

$$
\begin{aligned}
& 2\left[\tilde{x}^T(t) K^T Y_1 + \tilde{x}^T(t-d(t)) K^T Y_2 + w^T(t) Y_3\right] \times \int_{t-d(t)}^{t} K\dot{\tilde{x}}(s)\, ds \\
&+ \int_{t-d(t)}^{t} \dot{\tilde{x}}^T(s)\, e^{\alpha(s-t)} K^T Z_1 K \dot{\tilde{x}}(s)\, ds \\
&= \int_{t-d(t)}^{t} Y_{12}^T \left(e^{\alpha(s-t)} Z_1\right)^{-1} Y_{12} ds \\
&- \int_{t-d(t)}^{t} \varsigma^T(t) Y \left(e^{\alpha(s-t)} Z_1\right)^{-1} Y^T \varsigma(t)\, ds \\
&\geq -H_{20} \varsigma^T(t) Y Z_1^{-1} Y^T \varsigma(t), \qquad (5.19)
\end{aligned}
$$

where

$$
\begin{aligned}
& \varsigma^T(t) = \left[\, x^T(t)\ \hat{x}^T(t)\ \tilde{x}^T(t-d(t)) K^T\ w^T(t)\ \tilde{x}^T(t-h_1) K^T\ \tilde{x}^T(t-h_2) K^T \,\right], \\
& Y_{12}^T = \varsigma^T(t) Y + \dot{\tilde{x}}^T(s)\, e^{\alpha(s-t)} K^T Z_1, \\
& Y^T = \left[\, Y_1^T\ 0\ Y_2^T\ Y_3^T\ 0\ 0 \,\right].
\end{aligned}
$$

By using the same methods as (5.19), from (5.15)-(5.18), we get

$$
\begin{aligned}
& \int_{t-h_2}^{t-d(t)} \dot{\tilde{x}}^T(s)\, e^{\alpha(s-t)} K^T Z_1 K \dot{\tilde{x}}(s)\, ds \\
&+ \int_{t-h_2}^{t-d(t)} \dot{\tilde{x}}^T(s)\, e^{\alpha(s-t)} K^T Z_2 K \dot{\tilde{x}}(s)\, ds \\
&+ 2\left[\tilde{x}^T(t) K^T T_1 + \tilde{x}^T(t-d(t)) K^T T_2 + w^T(t) T_3\right] \times \int_{t-h_2}^{t-d(t)} K\dot{\tilde{x}}(s)\, ds \\
&\geq -H_{21} \varsigma^T(t) T (Z_1 + Z_2)^{-1} T^T \varsigma(t), \qquad (5.20)
\end{aligned}
$$

and

$$
\begin{aligned}
& 2\left[\tilde{x}^T(t) K^T L_1 + \tilde{x}^T(t-d(t)) K^T L_2 + w^T(t) L_3\right] \int_{t-d(t)}^{t-h_1} K\dot{\tilde{x}}(s)\, ds \\
&+ \int_{t-d(t)}^{t-h_1} \dot{\tilde{x}}^T(s)\, e^{\alpha(s-t)} K^T Z_2 K \dot{\tilde{x}}(s)\, ds \\
&\geq -H_{21} \varsigma^T(t) L Z_2^{-1} L^T \varsigma(t), \qquad (5.21)
\end{aligned}
$$

where

$$T^T = \begin{bmatrix} T_1^T & 0 & T_2^T & T_3^T & 0 & 0 \end{bmatrix},$$
$$L^T = \begin{bmatrix} L_1^T & 0 & L_2^T & L_3^T & 0 & 0 \end{bmatrix}.$$

Since the time delay satisfies (A1), so

$$\begin{aligned} &-\left(1-\dot{d}(t)\right)\tilde{x}^T(t-d(t))\,e^{-\alpha d(t)}K^T Q_3 K\tilde{x}(t-d(t)) \\ &\leq -(1-d)\,\tilde{x}^T(t-d(t))\,e^{-\alpha h_1}K^T Q_3 K\tilde{x}(t-d(t)). \end{aligned} \tag{5.22}$$

Without loss of generality, we partition P as

$$P = \begin{bmatrix} P_1 & -P_2 \\ * & P_2 \end{bmatrix}, \tag{5.23}$$

where $P_1 > 0$ and $P_2 > 0$. Since $P > 0$, by Schur Complement Lemma, we obtain (5.10).

From (5.9), (5.13), (5.19)-(5.23), one can obtain that

$$\dot{V}(t) + \alpha V(t) + \Gamma(t) \leq \varsigma^T(t)\,\Theta\,\varsigma(t), \tag{5.24}$$

where

$$\begin{aligned} \Theta &= \Omega_1 + \Gamma_2^T U \Gamma_2 + \Gamma_3^T \Gamma_3 + H_{20} Y Z_1^{-1} Y^T + H_{21} T Z_3^{-1} T^T + H_{21} L Z_2^{-1} L^T \\ \Gamma_2 &= \begin{bmatrix} A & 0 & E & B & 0 & 0 \end{bmatrix}, \\ \Gamma_3 &= \begin{bmatrix} H - D_f C & -C_f & \tilde{G} & \tilde{M} & 0 & 0 \end{bmatrix}, \\ \Omega_1 &= \begin{bmatrix} \varphi_{11} & \varphi_{12} & \varphi_{13} & \varphi_{14} & L_1 & -T_1 \\ * & \varphi_{22} & \varphi_{23} & \varphi_{24} & 0 & 0 \\ * & * & \varphi_{33} & \varphi_{34} & L_2 & -T_2 \\ * & * & * & -\gamma^2 I & L_3 & -T_3 \\ * & * & * & * & -e^{-\alpha h_1} Q_1 & 0 \\ * & * & * & * & * & -e^{-\alpha h_2} Q_2 \end{bmatrix}. \end{aligned}$$

By Schur Complement Lemma, we can get that $\Theta < 0$ is equivalent to (5.11). Therefore, it is obvious that

$$\dot{V}(t) + \alpha V(t) + \Gamma(t) \leq 0. \tag{5.25}$$

Integrating (5.25) during $[t_0, t]$ gives (5.12), which completes the proof.

Remark 5.6. When estimating the upper bound of the derivative of Lyapunov functional, some useful terms are often ignored, which may lead to considerable conservativeness. In [40], however, a method is proposed to estimate the upper bound of the derivative of Lyapunov functional without ignoring some useful terms. Lemma 5.5 is established by the method proposed in [40]. Therefore, the criterion based on Lemma 5.5 is expected to be less conservative.

Next, the exponential H_∞ performance analysis of the filtering error system (5.5) is presented based on Lemma 5.5 as follows.

Lemma 5.7. *Let $\alpha > 0$, $\gamma > 0$, $h_2 > h_1 \geq 0$, d and $\mu \geq 1$ be given constants, the filtering error system (5.5) under (A1) is exponentially stable with an H_∞ prescribed attenuation level γ, if there exist symmetric matrices $P_{il} > 0$, $Z_{il} > 0$, $Q_{ik} > 0$, $P_{jl} > 0$, $Z_{jl} > 0$, $Q_{jk} > 0$ and matrices Y_{ik}, T_{ik}, L_{ik}, $k = 1,2,3, l = 1,2, \forall i, j \in \mathscr{P}$ such that the following inequalities hold,*

$$P_{i1} - P_{i2} > 0,\ P_{j1} - P_{j2} > 0, \tag{5.26}$$

$$P_{i2} \leq \mu P_{j2},\ P_{i1} - P_{i2} \leq \mu(P_{j1} - P_{j2}),\ Q_{ik} \leq \mu Q_{jk},\ Z_{il} \leq \mu Z_{jl}, \tag{5.27}$$

$$\begin{bmatrix} \Omega_i & R_i \\ * & -W_i \end{bmatrix} < 0, \tag{5.28}$$

and average dwell time satisfies

$$\tau_a > \tau_a^* = \frac{\ln \mu}{\alpha}, \tag{5.29}$$

where

$$\Omega_i = \begin{bmatrix} \varphi_{i11} & \varphi_{i12} & \varphi_{i13} & \varphi_{i14} \\ * & \varphi_{i22} & \varphi_{i23} & \varphi_{i24} \\ * & * & \varphi_{i33} & \varphi_{i34} \\ * & * & * & -\gamma^2 I \end{bmatrix},$$

$$R_i = \begin{bmatrix} L_{i1} & -T_{i1} & A_i^T U_i & H_i^T - C_i^T D_{if}^T & Y_{i1} & T_{i1} & L_{i1} \\ 0 & 0 & 0 & -C_{if}^T & 0 & 0 & 0 \\ L_{i2} & -T_{i2} & E_i^T U_i & \tilde{G}_i^T & Y_{i2} & T_{i2} & L_{i2} \\ L_{i3} & -T_{i3} & B_i^T U_i & \tilde{M}_i^T & Y_{i3} & T_{i3} & L_{i3} \end{bmatrix},$$

$$W_i = diag\left\{ e^{-\alpha h_1} Q_{i1},\ e^{-\alpha h_2} Q_{i2},\ U_i,\ I,\ H_{20}^{-1} Z_{i1},\ H_{21}^{-1} Z_{i3},\ H_{21}^{-1} Z_{i2} \right\},$$

$$H_{21} = \frac{e^{\alpha h_2} - e^{\alpha h_1}}{\alpha},\ H_{20} = \frac{e^{\alpha h_2} - 1}{\alpha},\ h_{21} = h_2 - h_1,$$

$$U_i = h_2 Z_{i1} + h_{21} Z_{i2},\ Z_{i3} = Z_{i1} + Z_{i2},$$

$$\varphi_{i11} = P_{i1} A_i - P_{i2} B_{if} C_i + A_i^T P_{i1} - C_i^T B_{if}^T P_{i2} + \sum_{l=1}^{3} Q_{il} + Y_{i1} + Y_{i1}^T + \alpha P_{i1},$$

$$\varphi_{i12} = -P_{i2} A_{if} - A_i^T P_{i2} + C_i^T B_{if}^T P_{i2} - \alpha P_{i2},$$

$$\varphi_{i22} = P_{i2} A_{if} + A_{if}^T P_{i2} + \alpha P_{i2},$$

$$\varphi_{i13} = P_{i1} E_i - P_{i2} B_{if} F_i - Y_{i1} + Y_{i2}^T + T_{i1} - L_{i1},$$

$$\varphi_{i23} = -P_{i2} E_i + P_{i2} B_{if} F_i,$$

$$\varphi_{i33} = (d-1) e^{-\alpha h_1} Q_{i3} - Y_{i2} - Y_{i2}^T + T_{i2} + T_{i2}^T - L_{i2} - L_{i2}^T,$$

$$\varphi_{i14} = P_{i1} B_i - P_{i2} B_{if} D_i + Y_{i3}^T,$$

$$\varphi_{i24} = -P_{i2} B_i + P_{i2} B_{if} D_i,$$

$$\varphi_{i34} = T_{i3}^T - Y_{i3}^T - L_{i3}^T.$$

Proof. For the ith ($\sigma(t_k) = ith$) subsystems of the filtering error system (5.5), the corresponding Lyapunov function candidate is selected as

$$\begin{aligned}
V_i(t) &= \tilde{x}^T(t) P_i \tilde{x}(t) + e^{-\alpha t} \sum_{l=1}^{2} \int_{t-h_l}^{t} \tilde{x}^T(s) e^{\alpha s} K^T Q_{il} K \tilde{x}(s) ds \\
&+ e^{-\alpha t} \int_{t-d(t)}^{t} \tilde{x}^T(s) e^{\alpha s} K^T Q_{i3} K \tilde{x}(s) ds \\
&+ e^{-\alpha t} \int_{-h_2}^{0} \int_{t+\theta}^{t} \dot{\tilde{x}}^T(s) e^{\alpha s} K^T Z_{i1} K \dot{\tilde{x}}(s) ds d\theta \\
&+ e^{-\alpha t} \int_{-h_2}^{-h_1} \int_{t+\theta}^{t} \dot{\tilde{x}}^T(s) e^{\alpha s} K^T Z_{i2} K \dot{\tilde{x}}(s) ds d\theta,
\end{aligned} \tag{5.30}$$

where $P_i = \begin{bmatrix} P_{i1} & -P_{i2} \\ * & P_{i2} \end{bmatrix} > 0$, $Z_{il} > 0$, $Q_{ik} > 0, k = 1,2,3, l = 1,2$ are symmetric matrices.

By Lemma 5.5, if (5.26) and (5.28) hold, it is obtained that

$$V_i(t) \le e^{-\alpha(t-t_k)} V_i(t_k) - \int_{t_k}^{t} e^{-\alpha(t-\tau)} \Gamma(\tau) d\tau. \tag{5.31}$$

It follows from (5.27) and (5.30) that

$$V_i(t) \le \mu V_j(t) \quad \forall i, j \in \mathscr{P}. \tag{5.32}$$

Combining (5.31) and (5.32) yields, at the switching instants $t_0 < t_1 < \cdots < t_k < t$ during $[t_0, t]$,

$$\begin{aligned}
V_{\sigma(t_k)}(t) &\le V_{\sigma(t_k)}(t_k) e^{-\alpha(t-t_k)} - \int_{t_k}^{t} e^{-\alpha(t-\tau)} \Gamma(\tau) d\tau \\
&\le \mu V_{\sigma(t_k^-)}(t_k^-) e^{-\alpha(t-t_k)} - \int_{t_k}^{t} e^{-\alpha(t-\tau)} \Gamma(\tau) d\tau \\
&\le \mu \left[V_{\sigma(t_{k-1})}(t_{k-1}) e^{-\alpha(t_k - t_{k-1})} - \int_{t_{k-1}}^{t_k} e^{-\alpha(t_k-\tau)} \Gamma(\tau) d\tau \right] e^{\alpha(t_k - t)} \\
&- \int_{t_k}^{t} e^{-\alpha(t-\tau)} \Gamma(\tau) d\tau \le \cdots \le \mu^k e^{-\alpha(t-t_0)} V_{\sigma(t_0)}(t_0) \\
&- \mu^k \int_{t_0}^{t_1} e^{-\alpha(t-\tau)} \Gamma(\tau) d\tau - \cdots - \mu^0 \int_{t_k}^{t} e^{-\alpha(t-\tau)} \Gamma(\tau) d\tau \\
&= e^{-\alpha(t-t_0) + N_\sigma(t_0,t) \ln \mu} V_{\sigma(t_0)}(t_0) \\
&- \int_{t_0}^{t} e^{-\alpha(t-\tau) + N_\sigma(\tau,t) \ln \mu} \Gamma(\tau) d\tau.
\end{aligned} \tag{5.33}$$

Notice that (5.7), (5.29) and (5.33), we get

$$V_{\sigma(t_k)}(t) \le e^{-\lambda(t-t_0)} V_{\sigma(t_0)}(t_0) - \int_{t_0}^{t} e^{-\lambda(t-\tau)} \Gamma(\tau) d\tau, \tag{5.34}$$

where $\lambda = \alpha - \ln\mu/\tau_a$.

Assuming the zero disturbance input to the filtering error system (5.5) and $\Gamma(\tau) = 0$, it follows from (5.34) that

$$V_{\sigma(t_k)}(t) \le e^{-\lambda(t-t_0)} V_{\sigma(t_0)}(t_0). \tag{5.35}$$

According to (5.30), it is obtained that

$$\begin{aligned} &a_{\sigma(t_k)} \|\tilde{x}(t)\|^2 \le V_{\sigma(t_k)}(t), \\ &V_{\sigma(t_0)}(t_0) \le b_{\sigma(t_0)} \|\tilde{x}(t_0)\|_\theta^2. \end{aligned} \tag{5.36}$$

Combining (5.35) and (5.36) yields

$$\|\tilde{x}(t)\| \le \sqrt{\frac{b_{\sigma(t_0)}}{a_{\sigma(t_k)}}} e^{-\lambda(t-t_0)/2} \|\tilde{x}(t_0)\|_\theta, \tag{5.37}$$

where

$$\begin{aligned} a_{\sigma(t_k)} &= \lambda_{\min}\left(P_{\sigma(t_k)}\right), \\ b_{\sigma(t_0)} &= \lambda_{\max}\left(P_{\sigma(t_0)}\right) + \left(1 - e^{-\alpha h_1}\right) \lambda_{\max}\left(K^T Q_{1\sigma(t_0)} K\right)/\alpha \\ &\quad + \left(1 - e^{-\alpha h_2}\right) \left(\lambda_{\max}\left(K^T Q_{2\sigma(t_0)} K\right) + \lambda_{\max}\left(K^T Q_{3\sigma(t_0)} K\right)\right)/\alpha \\ &\quad + \left(e^{-\alpha h_2} + \alpha h_2 - 1\right) \lambda_{\max}\left(K^T Z_{1\sigma(t_0)} K\right)/\alpha^2 \\ &\quad + \left(e^{-\alpha h_2} + \alpha h_2 - e^{-\alpha h_1} - \alpha h_1\right) \lambda_{\max}\left(K^T Z_{2\sigma(t_0)} K\right)/\alpha^2. \end{aligned}$$

Then, the filtering error system (5.5) is exponentially stable.

Now, to establish the H_∞ performance for the filtering error system (5.5), consider the following performance index

$$J = \int_0^\infty \left(\tilde{z}^T(t)\tilde{z}(t) - \gamma^2 w^T(t) w(t)\right) dt = \int_0^\infty \Gamma(t)\, dt.$$

From (5.34), let $t_0 = 0$ and under zero initial condition $V_i(\tilde{x}(t_0))|_{t_0=0} = 0$, we have

$$\int_0^t e^{-\lambda(t-\tau)} \Gamma(\tau)\, d\tau \le 0. \tag{5.38}$$

Note in the above inequality that

$$\int_0^t e^{-\lambda(t-\tau)} \tilde{z}^T(\tau)\tilde{z}(\tau)\, d\tau,$$

is integrable from $t = 0$ to ∞ since

$$\int_0^t e^{-\lambda(t-\tau)} w^T(\tau) w(\tau) d\tau,$$

is integrable for any $w(t) \in l_2[0,\infty)$ in the same interval. Then, we integrate (5.38) from $t = 0$ to ∞ to obtain that

$$\int_0^\infty \left(\int_0^t e^{-\lambda(t-\tau)} \Gamma(\tau) d\tau \right) dt \leq 0. \tag{5.39}$$

Rearranging the double-integral area yields

$$\begin{aligned} &\int_0^\infty \Gamma(\tau) \left(\int_\tau^\infty e^{-\lambda(t-\tau)} dt \right) d\tau \\ &= \frac{1}{\lambda} \int_0^\infty \Gamma(\tau) d\tau \leq 0, \end{aligned} \tag{5.40}$$

which means that $J \leq 0$. Then, one has $\|\tilde{z}(t)\|_2 \leq \gamma \|w(t)\|_2$ for any $w(t) \in l_2[0,\infty)$. The proof is completed.

Remark 5.8. If $\mu = 1$, which leads to $P_{il} \equiv P_l > 0$, $Z_{il} \equiv Z_l > 0$, $Q_{ik} \equiv Q_k > 0$, $k = 1,2,3, l = 1,2, \forall i \in \mathscr{P}$ by (5.27) and $\tau^* = 0$ by (5.29), then the filtering error system (5.5) possesses a common Lyapunov functional and the switching signals can be arbitrary. This implies that system (5.5) under the arbitrary switching signals is exponentially stable with a prescribed H_∞ attenuation level γ.

Remark 5.9. From (5.37), we can know that the filtering error system (5.5) with $w(t) = 0$ is exponentially stable. In fact, in practical applications, the time-varying delay often lies in a range, that is, the lower bound of the time-varying delay is not zero, which is difficult to deal with. In this case, our results are applicable.

5.1.3 H_∞ Filter Design

In this section, the following theorem provides sufficient conditions for the existence of a filter in form of (5.4) assuring an exponential H_∞ performance for system (5.1) and shows how to determine the filter parameter matrices based on Lemma 5.7.

Theorem 5.10. *Consider system (5.1) with (A1) and let $\alpha > 0$, $\gamma > 0$, $h_2 > h_1 \geq 0$, d and $\mu \geq 1$ be given constants. If there exist symmetric and positive definite matrices P_{il}, Z_{il}, Q_{ik}, P_{jl}, Z_{jl}, Q_{jk} and matrices Y_{ik}, T_{ik}, L_{ik}, N_{i1}, N_{i2}, N_{i3}, N_{i4}, $k = 1,2,3, l = 1,2, \forall i, j \in \mathscr{P}$ such that (5.26), (5.27), (5.29) and the following inequalities hold,*

$$\begin{bmatrix} \tilde{\Omega}_i & \tilde{R}_i \\ * & -W_i \end{bmatrix} < 0, \tag{5.41}$$

where

$$\tilde{\Omega}_i = \begin{bmatrix} \tilde{\varphi}_{i11} & \tilde{\varphi}_{i12} & \tilde{\varphi}_{i13} & \tilde{\varphi}_{i14} \\ * & \tilde{\varphi}_{i22} & \tilde{\varphi}_{i23} & \tilde{\varphi}_{i24} \\ * & * & \varphi_{i33} & \varphi_{i34} \\ * & * & * & -\gamma^2 I \end{bmatrix},$$

$$\tilde{R}_i = \begin{bmatrix} L_{i1} & -T_{i1} & A_i^T U_i & H_i^T - C_i^T N_{i4}^T & Y_{i1} & T_{i1} & L_{i1} \\ 0 & 0 & 0 & -N_{i3}^T & 0 & 0 & 0 \\ L_{i2} & -T_{i2} & E_i^T U_i & G_i^T - F_i^T N_{i4}^T & Y_{i2} & T_{i2} & L_{i2} \\ L_{i3} & -T_{i3} & B_i^T U_i & M_i^T - D_i^T N_{i4}^T & Y_{i3} & T_{i3} & L_{i3} \end{bmatrix},$$

$$\tilde{\varphi}_{i11} = P_{i1}A_i - N_{i2}C_i + A_i^T P_{i1} - C_i^T N_{i2}^T + \sum_{l=1}^{3} Q_{il} + Y_{i1} + Y_{i1}^T + \alpha P_{i1},$$

$$\tilde{\varphi}_{i12} = -N_{i1} - A_i^T P_{i2} + C_i^T N_{i2}^T - \alpha P_{i2},$$

$$\tilde{\varphi}_{i22} = N_{i1} + N_{i1}^T + \alpha P_{i2},$$

$$\tilde{\varphi}_{i13} = P_{i1}E_i - N_{i2}F_i - Y_{i1} + Y_{i2}^T + T_{i1} - L_{i1},$$

$$\tilde{\varphi}_{i23} = -P_{i2}E_i + N_{i2}F_i,$$

$$\tilde{\varphi}_{i14} = P_{i1}B_i - N_{i2}D_i + Y_{i3}^T,$$

$$\tilde{\varphi}_{i24} = -P_{i2}B_i + N_{i2}D_i,$$

and W_i, U_i, φ_{i33} and φ_{i34} are defined in Lemma 5.7.

Then, there exists a filter in form of (5.4) such that the corresponding filtering error system (5.5) under the switching signals with average dwell time satisfying (5.29) is exponentially stable with an H_∞ prescribed attenuation level γ. Moreover, if LMIs (5.26), (5.27) and (5.41) have a feasible solution, the filter matrices can be constructed by

$$A_{if} = P_{i2}^{-1}N_{i1},\ B_{if} = P_{i2}^{-1}N_{i2},\ C_{if} = N_{i3},\ D_{if} = N_{i4}. \tag{5.42}$$

Proof. By Lemma 5.7, the filtering error system (5.5) is exponentially stable with an H_∞ prescribed attenuation level γ, if LMIs (5.26), (5.27) and (5.28) hold and (5.29) is satisfied. Define

$$N_{i1} = P_{i2}A_{if},\ N_{i2} = P_{i2}B_{if},\ N_{i3} = C_{if},\ N_{i4} = D_{if}. \tag{5.43}$$

By substituting (5.43) into (5.41), it can be easily established that (5.41) is equivalent to (5.28). Thus, if LMIs (5.26), (5.27) and (5.41) hold and (5.29) is satisfied, the filtering error system (5.5) is exponentially stable with an H_∞ prescribed attenuation level γ. Furthermore, if a solution to LMIs (5.26), (5.27) and (5.41) exists, the parameters of an admissible filter are given by (5.42). The proof is completed.

Remark 5.11. It should be mentioned that the interval time-varying delay in the above criteria satisfies the case (A1). However, in many cases, the information of the derivative of time delay is unknown. As for this circumstance, i.e. (A2), these criteria in this paper are derived provided that the terms containing Q_3 in Lemma 5.5 and Q_{i3} in Lemma 5.7 and Theorem 5.10 are removed. Therefore, sufficient conditions for the existence of a filter in the form of (5.4) assuring an exponential H_∞ performance for system (5.1) with (A2) are obtained as follows.

Corollary 5.12. *Consider system (5.1) with (A2) and let $\alpha > 0$, $\gamma > 0$, $h_2 > h_1 \geq 0$ and $\mu \geq 1$ be given constants. If there exist symmetric and positive definite matrices P_{il}, Z_{il}, Q_{il}, P_{jl}, Z_{jl}, Q_{jl} and matrices Y_{ik}, T_{ik}, L_{ik}, N_{i1}, N_{i2}, N_{i3}, N_{i4}, $l = 1,2, \forall i, j \in \mathscr{P}$ such that (5.26), (5.29) and the following inequalities hold,*

$$P_{i2} \leq \mu P_{j2},\ P_{i1} - P_{i2} \leq \mu(P_{j1} - P_{j2}),\ Q_{il} \leq \mu Q_{jl},\ Z_{il} \leq \mu Z_{jl}, \tag{5.44}$$

$$\begin{bmatrix} \hat{\Omega}_i & \tilde{R}_i \\ * & -W_i \end{bmatrix} < 0, \tag{5.45}$$

where

$$\hat{\Omega}_i = \begin{bmatrix} \hat{\varphi}_{i11} & \tilde{\varphi}_{i12} & \tilde{\varphi}_{i13} & \tilde{\varphi}_{i14} \\ * & \tilde{\varphi}_{i22} & \tilde{\varphi}_{i23} & \tilde{\varphi}_{i24} \\ * & * & \hat{\varphi}_{i33} & \varphi_{i34} \\ * & * & * & -\gamma^2 I \end{bmatrix},$$

$$\hat{\varphi}_{i11} = P_{i1}A_i - N_{i2}C_i + A_i^T P_{i1} - C_i^T N_{i2}^T + \sum_{l=1}^{2} Q_{il} + Y_{i1} + Y_{i1}^T + \alpha P_{i1},$$

$$\hat{\varphi}_{i33} = -Y_{i2} - Y_{i2}^T + T_{i2} + T_{i2}^T - L_{i2} - L_{i2}^T,$$

and $\tilde{R}_i$, W_i, $\tilde{\varphi}_{i12}$, $\tilde{\varphi}_{i22}$, $\tilde{\varphi}_{i13}$, $\tilde{\varphi}_{i23}$, $\tilde{\varphi}_{i14}$, $\tilde{\varphi}_{i24}$, φ_{i34} are defined in Theorem 5.10.

Then, there exists a filter in form of (5.4) such that the corresponding filtering error system (5.5) under the switching signals with average dwell time satisfying (5.29) is exponentially stable with an H_∞ prescribed attenuation level γ. Moreover, if LMIs (5.26), (5.44) and (5.45) have a feasible solution, the filter matrices can be constructed by (5.42).

Remark 5.13. There are two methods to obtain μ: one is that we first select a scalar as μ, then solve some LMIs. If these LMIs have no solution, a larger μ will be selected to yield a solution. On the contrary, a smaller μ will be chosen. So, a proper μ could be obtained. The other is that we first solve some LMIs, then find μ according to the obtained solution. This method may not lead to a suitable μ. Therefore, we utilize the former to obtain μ in this paper.

Remark 5.14. With the result of Theorem 5.10, the exponential H_∞ filtering problem for system (5.1) with (A1) can be solved by testing the feasibility of LMIs (5.26), (5.27) and (5.41). Note that any feasible solution to LMIs (5.26), (5.27) and (5.41) leads to a suitable filter by means of (5.42). In addition, it follows from Theorem 5.10 that the minimum noise attenuation level γ^* can be obtained from the solution of the following problem.

Problem 5.15.

$$\begin{aligned}&Min\ \gamma,\\&\text{subject to } (5.26),\ (5.27)\ and\ (5.41)\\&\text{with } P_{il},\ Z_{il},\ Q_{ik},\ P_{jl},\ Z_{jl},\ Q_{jk},\ Y_{ik},\ T_{ik},\ L_{ik},\ N_{i1},\ N_{i2},\ N_{i3},\ N_{i4}.\end{aligned}$$

The corresponding filter matrices are given by (5.42). Regarding (A2), the exponential H_∞ filtering problem for system (5.1) can be solved by testing the feasibility of LMIs (5.26), (5.44) and (5.45) and a desired filter can be developed by (5.42). Moreover, the noise attenuation level γ can be optimized by the following optimization problem as follows.

Problem 5.16.

$$\begin{aligned}&\gamma^* = Min\ \gamma,\\&\text{subject to } (5.26),\ (5.44)\ and\ (5.45)\\&\text{with } P_{il},\ Z_{il},\ Q_{il},\ P_{jl},\ Z_{jl},\ Q_{jl},\ Y_{ik},\ T_{ik},\ L_{ik},\ N_{i1},\ N_{i2},\ N_{i3},\ N_{i4}.\end{aligned}$$

The corresponding filter matrices are given by (5.42).

5.2 Delay-Dependent H_∞ Model Reduction

As pointed out in Chapter 4, high-order mathematical models lead to great difficulties to analysis and synthesis of the systems and model reduction issues have attracted great attention from researchers and engineers in the system and control community during the past few years. In the discrete-time context, the model reduction problem has been studied in the previous chapter. In the continuous-time context, sufficient conditions for the solvability of this problem are presented for time-delay systems [133], singular systems [130], Markovian jump systems [143], linear parameter-varying systems with distributed delay [121] and switched delay systems [122]. However, all the conditions obtained in [133, 130, 143, 121, 122] are formulated in terms of LMIs with some non-convex constraints. These non-convex conditions make LMIs difficult to find the numerical solution, which is NP-hard [143]. The algorithms to solve such LMIs cause large amount of calculation and sometimes fail to guarantee the global convergence. Therefore, it is desired to derive the strict LMIs formulation. In this section, by introducing a block-symmetric matrix and using an ADT approach, sufficient conditions for the existence of the desired reduced-order model under ADT switching signals are derived and formulated in terms of strict LMIs.

5.2.1 Problem Formulation

Consider a class of switched systems with interval time-varying delay:

$$\begin{aligned} \dot{x}(t) &= A_\sigma x(t) + E_\sigma x(t-d(t)) + B_\sigma u(t), \\ y(t) &= C_\sigma x(t) + F_\sigma x(t-d(t)) + D_\sigma u(t), \\ x(\theta) &= \phi(\theta), \quad \theta \in [-h_2, 0] \end{aligned} \tag{5.46}$$

where $x(t) \in R^n$ is the state, $y(t) \in R^s$ is the measured output, $u(t) \in R^r$ is the input vector which is assumed to belong to $l_2[0,\infty)$. The piecewise constant function σ denoting $\sigma(t)$ for simplicity : $[0,\ \infty) \to \mathscr{P} = \{1, \cdots, p\}$ is a switching signal to specify, at each time instant t, the index $\sigma \in \mathscr{P}$ of the active subsystem, i.e., $\sigma = i$ means that the ith subsystem is activated, $p > 1$ is the number of subsystems. As in [90] and the previous chaters, it is assumed that the switching signal is not known a priori, but its instantaneous value is available in real-time implementation. The ith subsystem is denoted by constant matrices A_i, E_i, B_i, C_i, F_i and D_i with the appropriate dimensions. $d(t)$ is assumed to be the interval time-varying delay satisfying either of two conditions (A1) and (A2).

In this section, a desired reduced-order model that we will design is described by

$$\begin{aligned} \dot{\hat{x}}(t) &= A_{ri}\hat{x}(t) + E_{ri}\hat{x}(t-d(t)) + B_{ri}u(t), \\ \hat{y}(t) &= C_{ri}\hat{x}(t) + F_{ri}\hat{x}(t-d(t)) + D_{ri}u(t), \end{aligned} \tag{5.47}$$

where $\hat{x}(t) \in R^q$ is the state vector of the reduced-order system with $q < n$, $\hat{y}(t) \in R^s$ is the output vector of the reduced-order system with the same dimension s as the original output $y(t)$, and A_{ri}, B_{ri}, C_{ri}, D_{ri}, E_{ri} and F_{ri}, $\forall i \in \mathscr{P}$ are the matrices to be determined. It is assumed that the switching signal σ in system (5.47) is available in real-time and simultaneous with the one in system (5.46).

Augmenting the model of system (5.46) to include the state of system (5.47), we can obtain the following error system

$$\begin{aligned} \dot{\tilde{x}}(t) &= \tilde{A}_i\tilde{x}(t) + \tilde{E}_i\tilde{x}(t-d(t)) + \tilde{B}_iu(t), \\ \tilde{e}(t) &= \tilde{H}_i\tilde{x}(t) + \tilde{G}_i\tilde{x}(t-d(t)) + \tilde{M}_iu(t), \end{aligned} \tag{5.48}$$

where

$$\begin{aligned} &\tilde{x}(t) = \left[x^T(t)\ \hat{x}^T(t)\right]^T, \ \tilde{e}(t) = y(t) - \hat{y}(t), \\ &\tilde{A}_i = \begin{bmatrix} A_i & 0 \\ 0 & A_{ri} \end{bmatrix}, \ \tilde{E}_i = \begin{bmatrix} E_i & 0 \\ 0 & E_{ri} \end{bmatrix}, \\ &\tilde{B}_i = \left[B_i^T\ B_{ri}{}^T\right]^T, \ \tilde{H}_i = \left[C_i\ -C_{ri}\right], \\ &\tilde{G}_i = \left[F_i\ -F_{ri}\right], \ \tilde{M}_i = D_i - D_{ri}. \end{aligned}$$

Our objective is to design a reduced-order system in the form of (5.47), and find admissible ADT switching signals such that system (5.48) is exponentially stable with an H_∞ norm bound.

Before ending this section, we present the following lemmas which will play an important role in our further derivation.

Lemma 5.17. *For the case (A1), let $\alpha > 0$, $\gamma > 0$, $h_2 > h_1 \geq 0$, d and $\mu \geq 1$ be given constants, the error system (5.48) under ADT switching signals is exponentially stable with an H_∞ norm bound γ, if there exist symmetric and positive definite matrices P_i, Q_{ik}, Z_{il} and matrices Y_{ik}, T_{ik}, L_{ik}, $k = 1,2,3, l = 1,2, \forall i \in \mathscr{P}$ such that the following inequalities hold,*

$$\begin{bmatrix} \Omega_i & R_i \\ * & -W_i \end{bmatrix} < 0, \tag{5.49}$$

$$P_i \leq \mu P_j,\ Q_{ik} \leq \mu Q_{jk},\ Z_{il} \leq \mu Z_{jl},\ \forall i,\ j \in \mathscr{P}, \tag{5.50}$$

and average dwell time τ_a satisfies

$$\tau_a > \frac{\ln \mu}{\alpha}, \tag{5.51}$$

where

$$\Omega_i = \begin{bmatrix} \varphi_{i11} & \varphi_{i12} & \varphi_{i13} \\ * & \varphi_{i22} & \varphi_{i23} \\ * & * & -\gamma^2 I \end{bmatrix},$$

$$\begin{aligned}
\varphi_{i11} &= P_i\tilde{A}_i + \tilde{A}_i^T P_i + \sum_{l=1}^{3} Q_{il} + Y_{i1} + Y_{i1}^T + \alpha P_i, \\
\varphi_{i12} &= P_i\tilde{E}_i - Y_{i1} + Y_{i2}^T + T_{i1} - L_{i1}, \\
\varphi_{i13} &= P_i\tilde{B}_i + Y_{i3}^T, \\
\varphi_{i22} &= (d-1)\,e^{-\alpha h_1} Q_{i3} - Y_{i2} - Y_{i2}^T + T_{i2} + T_{i2}^T - L_{i2} - L_{i2}^T, \\
\varphi_{i23} &= T_{i3}^T - Y_{i3}^T - L_{i3}^T, \\
R_i &= \begin{bmatrix} L_{i1} & -T_{i1} & \tilde{A}_i^T & \tilde{H}_i^T & Y_{i1} & T_{i1} & L_{i1} \\ L_{i2} & -T_{i2} & \tilde{E}_i^T & \tilde{G}_i^T & Y_{i2} & T_{i2} & L_{i2} \\ L_{i3} & -T_{i3} & \tilde{B}_i^T & \tilde{M}_i^T & Y_{i3} & T_{i3} & L_{i3} \end{bmatrix}, \\
W_i &= diag\left\{ e^{-\alpha h_1} Q_{i1},\ e^{-\alpha h_2} Q_{i2},\ U_i^{-1},\ I,\ H_{20}^{-1} Z_{i1},\ H_{21}^{-1} Z_{i3},\ H_{21}^{-1} Z_{i2} \right\}, \\
H_{21} &= \frac{e^{\alpha h_2} - e^{\alpha h_1}}{\alpha},\ H_{20} = \frac{e^{\alpha h_2} - 1}{\alpha}, \\
h_{21} &= h_2 - h_1,\ U_i = h_2 Z_{i1} + h_{21} Z_{i2},\ Z_{i3} = Z_{i1} + Z_{i2}.
\end{aligned}$$

Proof. For the ith subsystems of the error system (5.48), the corresponding Lyapunov functional candidate is chosen as

$$\begin{aligned}V_i(t) &= \tilde{x}^T(t)P_i\tilde{x}(t)\\ &+\sum_{l=1}^{2}\int_{t-h_l}^{t}\tilde{x}^T(s)e^{\alpha(s-t)}Q_{il}\tilde{x}(s)ds\\ &+\int_{t-d(t)}^{t}\tilde{x}^T(s)e^{\alpha(s-t)}Q_{i3}\tilde{x}(s)ds\\ &+\int_{-h_2}^{0}\int_{t+\theta}^{t}\dot{\tilde{x}}^T(s)e^{\alpha(s-t)}Z_{i1}\dot{\tilde{x}}(s)dsd\theta\\ &+\int_{-h_2}^{-h_1}\int_{t+\theta}^{t}\dot{\tilde{x}}^T(s)e^{\alpha(s-t)}Z_{i2}\dot{\tilde{x}}(s)dsd\theta,\end{aligned}\tag{5.52}$$

where $P_i>0$, $Z_{il}>0$, $Q_{ik}>0, k=1,2,3, l=1,2$ are symmetric matrices. Based on (5.52), the proof is similar to that of Lemmas 5.7 and thus is omitted. □

Lemma 5.18. *For the case (A2), let $\alpha>0$, $h_2>h_1\geq 0$, $\gamma>0$ and $\mu\geq 1$ be given constants, the error system (5.48) under ADT switching signals is exponentially stable with an H_∞ norm bound γ, if there exist symmetric and positive definite matrices P_i, Q_{il}, Z_{il} and matrices Y_{ik}, T_{ik}, L_{ik}, $k=1,2,3, l=1,2, \forall i\in\mathscr{P}$ such that ADT satisfies (5.51) and the following inequalities hold,*

$$\begin{bmatrix}\hat{\Omega}_i & R_i\\ * & -W_i\end{bmatrix}<0,\tag{5.53}$$

$$P_i\leq\mu P_j,\ Q_{il}\leq\mu Q_{jl},\ Z_{il}\leq\mu Z_{jl},\ \forall i,\ j\in\mathscr{P},\tag{5.54}$$

where

$$\Omega_i=\begin{bmatrix}\hat{\varphi}_{i11} & \varphi_{i12} & \varphi_{i13}\\ * & \hat{\varphi}_{i22} & \varphi_{i23}\\ * & * & -\gamma^2 I\end{bmatrix},$$

$$\hat{\varphi}_{i11}=P_i\tilde{A}_i+\tilde{A}_i^TP_i+\sum_{l=1}^{2}Q_{il}+Y_{i1}+Y_{i1}^T+\alpha P_i,$$

$$\hat{\varphi}_{i22}=-Y_{i2}-Y_{i2}^T+T_{i2}+T_{i2}^T-L_{i2}-L_{i2}^T,$$

and R_i, W_i, φ_{i12}, φ_{i22}, φ_{i13}, φ_{i23} are defined in Lemma 5.17.

Proof. Since the information of the derivative of time delay is unknown, based on the condition in Lemma 5.17, Lemma 5.18 is obtained by virtue of removing the terms containing Q_{i3} in Lemma 5.17.

5.2.2 Design of H_∞ Reduced-Order Model

In this section, sufficient conditions for the existence of an exponential H_∞ reduced-order model (5.47) are presented. We first focus our attention on the case (A1). Secondly, the case (A2) is studied.

Theorem 5.19. *Consider system (5.46) with (A1) and let $\alpha > 0$, $\gamma > 0$, $h_2 > h_1 \geq 0$, d and $\mu \geq 1$ be given constants, if there exist positive definite and symmetric matrices $P_{i1} \in R^{n\times n}$, $P_{i2} \in R^{q\times q}$, $Z_{il} \in R^{m\times m}$, $Q_{ik} \in R^{m\times m}$ and matrices $Y_{ik} \in R^{m\times m}$, $T_{ik} \in R^{m\times m}$, $L_{ik} \in R^{m\times m}$, $N_{i1} \in R^{q\times q}$, $N_{i2} \in R^{q\times r}$, $N_{i3} \in R^{s\times q}$, $N_{i4} \in R^{s\times r}$, $N_{i5} \in R^{s\times q}$, $N_{i6} \in R^{s\times r}$, $q < n, m = n+q$, $k = 1,2,3, l = 1,2, \forall i \in \mathscr{P}$ such that (5.51) is satisfied and the following inequalities hold,*

$$\begin{bmatrix} \tilde{\Omega}_i & \tilde{R}_i \\ * & -\tilde{W}_i \end{bmatrix} < 0, \tag{5.55}$$

$$P_{i1} - EP_{i2}E^T > 0, \tag{5.56}$$

$$\Phi_i < \mu\Phi_j,\ Q_{ik} \leq \mu Q_{jk},\ Z_{il} \leq \mu Z_{jl},\ \forall i,j \in \mathscr{P}. \tag{5.57}$$

where

$$\tilde{\Omega}_i = \begin{bmatrix} \tilde{\varphi}_{i11} & \tilde{\varphi}_{i12} & \tilde{\varphi}_{i13} \\ * & \varphi_{i22} & \varphi_{i23} \\ * & * & -\gamma^2 I \end{bmatrix},$$

$$\tilde{\varphi}_{i11} = \tilde{\varphi}_{i16} + \tilde{\varphi}_{i16}^T + \sum_{l=1}^{3} Q_{il} + Y_{i1} + Y_{i1}^T + \alpha\Phi_i,$$

$$\Phi_i = \begin{bmatrix} P_{i1} & EP_{i2} \\ P_{i2}E^T & P_{i2} \end{bmatrix},\ E = \begin{bmatrix} I_{q\times q} \\ 0_{(n-q)\times q} \end{bmatrix},$$

$$\tilde{\varphi}_{i12} = \tilde{\varphi}_{i26} - Y_{i1} + Y_{i2}^T + T_{i1} - L_{i1},$$

$$\tilde{\varphi}_{i13} = \tilde{\varphi}_{i36} + Y_{i3}^T,$$

$$\tilde{R}_i = \begin{bmatrix} L_{i1} & -T_{i1} & \tilde{\varphi}_{i16}^T & \tilde{\varphi}_{i17}^T & Y_{i1} & T_{i1} & L_{i1} \\ L_{i2} & -T_{i2} & \tilde{\varphi}_{i26}^T & \tilde{\varphi}_{i27}^T & T_{i2} & L_{i2} & \\ L_{i3} & -T_{i3} & \tilde{\varphi}_{i36}^T & \tilde{\varphi}_{i37}^T & Y_{i3} & T_{i3} & L_{i3} \end{bmatrix},$$

$$\tilde{\varphi}_{i16} = \begin{bmatrix} P_{i1}A_i & EN_{i1} \\ P_{i2}E^T A_i & N_{i1} \end{bmatrix},$$

$$\tilde{\varphi}_{i26} = \begin{bmatrix} P_{i1}E_i & EN_{i2} \\ P_{i2}E^T E_i & N_{i2} \end{bmatrix},$$

$$\tilde{\varphi}_{i36} = \begin{bmatrix} P_{i1}B_i + EN_{i3} \\ P_{i2}E^T B_i + N_{i3} \end{bmatrix},$$

$$\tilde{\varphi}_{i17} = \begin{bmatrix} C_i & -N_{i4} \end{bmatrix},\ \tilde{\varphi}_{i27} = \begin{bmatrix} F_i & -N_{i5} \end{bmatrix},\ \tilde{\varphi}_{i37} = D_i - N_{i6},$$

$$\tilde{W}_i = diag\left\{ e^{-\alpha h_1}Q_{i1},\ e^{-\alpha h_2}Q_{i2},\ 2P_i - U_i,\ I,\ H_{20}^{-1}Z_{i1},\ H_{21}^{-1}Z_{i3},\ H_{21}^{-1}Z_{i2} \right\},$$

and H_{20}, H_{21}, U_i, φ_{i22} *and* φ_{i23} *are defined in Lemma 1. Then, there exists a q-order model (5.47) for an n-order original system (5.46) such that the error system (5.48) under ADT switching signals is exponentially stable with an* H_∞ *norm bound* γ. *Moreover, if LMIs (5.55), (5.56) and (5.57) have a feasible solution, switching signals with average dwell time* τ_a *satisfying (5.51) are found and an admissible q-order model (5.47) is constructed by*

$$\begin{aligned} &A_{ri} = P_{i2}^{-1}N_{i1},\ E_{ri} = P_{i2}^{-1}N_{i2},\ B_{ri} = P_{i2}^{-1}N_{i3},\\ &C_{ri} = N_{i4},\ F_{ri} = N_{i5},\ D_{ri} = N_{i6}. \end{aligned} \tag{5.58}$$

Proof. By Lemma 1, the error system (5.48) is exponentially stable with an H_∞ norm bound γ, if LMIs (5.49) and (5.50) hold, and (5.51) is satisfied.

From the fact that $(U_i - P_i)U_i^{-1}(U_i - P_i) \geq 0$, where $U_i > 0$ and $P_i > 0$, we have

$$-P_iU_i^{-1}P_i \leq U_i - 2P_i. \tag{5.59}$$

By using the inequality above, we can infer that if the following inequality holds,

$$\begin{bmatrix} \Omega_i & \bar{R}_i \\ * & -\tilde{W}_i \end{bmatrix} < 0, \tag{5.60}$$

where

$$\bar{R}_i = \begin{bmatrix} L_{i1} & -T_{i1} & \tilde{A}_i^T P_i & \tilde{H}_i^T & Y_{i1} & T_{i1} & L_{i1} \\ L_{i2} & -T_{i2} & \tilde{E}_i^T P_i & \tilde{G}_i^T & Y_{i2} & T_{i2} & L_{i2} \\ L_{i3} & -T_{i3} & \tilde{B}_i^T P_i & \tilde{M}_i^T & Y_{i3} & T_{i3} & L_{i3} \end{bmatrix},$$

then, the following inequality holds,

$$\begin{bmatrix} \Omega_i & \bar{R}_i \\ * & -\bar{W}_i \end{bmatrix} < 0, \tag{5.61}$$

where

$$\bar{W}_i = diag\left\{ e^{-\alpha h_1}Q_{i1},\ e^{-\alpha h_2}Q_{i2},\ P_iU_i^{-1}P_i,\ I,\ H_{20}^{-1}Z_{i1},\ H_{21}^{-1}Z_{i3},\ H_{21}^{-1}Z_{i2} \right\}.$$

Performing a congruence transformation to (5.61) via

$$diag\left\{I,\ I,\ I,\ I,\ I,\ P_i^{-1},\ I,\ I,\ I,\ I\right\},$$

yields (5.49). Partition

$$P_i = \Phi_i, \tag{5.62}$$

and let

$$\begin{aligned} &N_{i1} = P_{i2}A_{ri},\ N_{i2} = P_{i2}E_{ri},\ N_{i3} = P_{i2}B_{ri},\\ &N_{i4} = C_{ri}, N_{i5} = F_{ri},\ N_{i6} = D_{ri}. \end{aligned} \tag{5.63}$$

Substituting (5.63) into (5.60) yields (5.55), which implies that (5.49) holds. On the other hand, since P_i, P_{i1}, P_{i2} are positive definite and symmetric, one can get (5.56) by Schur complement. In addition, from (5.50) and (5.63), we can derive (5.57). Thus, if LMIs (5.55), (5.56) and (5.57) hold, the error system (5.48) under ADT switching signals is exponentially stable with an H_∞ norm bound γ. Furthermore, if a solution to LMIs (5.55), (5.56) and (5.57) exists, then switching signals with average dwell time τ_a are obtained by (5.51) and the parameters of an admissible reduced-order model are given by (5.58). The proof is completed.

Remark 5.20. It is worth pointing out that in the continuous-time context, many existing results on model reduction are expressed as LMIs with some non-convex constraints [143, 133, 130, 122]. Sufficient conditions for the existence of desired reduced-order models are obtained in terms of LMIs with a coupling rank constraint set in [133, 130]. The conditions are formulated as LMIs with matrix inverse constraints for Markovian jump linear systems [143] and switched delay systems under ADT switching signals [122]. These non-convex conditions make it difficult to find the numerical solutions, which is NP-hard [143]. This means that any algorithm for such conditions has exponential running time or fails to guarantee the convergence. All the results mentioned above are obtained by using Elimination Lemma [51], which leads to the conditions in terms of LMIs with non-convex constraints. In such cases, we avoid using Elimination Lemma. By virtue of (5.59) and (5.62), the strict LMI formulations are obtained in Theorem 1 for designing an admissible reduced-order models for continuous-time switched systems with interval time-varying delay under ADT switching signals, which are improved results established in [122].

Remark 5.21. Theorem 5.19 presents strict LMIs conditions for the existence of desired reduced-order models for continuous-time switched delay systems under ADT switching signals by resorting to (5.59) and (5.62). On one hand, we can see from (5.49) in Lemma 5.17 that it is difficult to deal with the problem of exponential H_∞ model reduction due to the existence of U_i^{-1}. (5.59) is exploited to overcome the difficulties. That is, U_i is not involved in any product with $\tilde{A}_i$, $\tilde{E}_i$ and $\tilde{B}_i$. On the other hand, to obtain strict LMIs-based conditions, in (5.62), P_i is partitioned as Φ_i in order to perform the variable replacement $N_{i1} = P_{i2}A_{ri}$. These two operations make more tractable to cope with the problem of exponential H_∞ model reduction in term of strict LMIs.

It is worth pointing out that the interval time-varying delay in Theorem 5.19 satisfies the case (A1). However, in many practical applications, the information of the derivative of time delay is unknown or difficult to get. In this case, the following theorem can be given.

Theorem 5.22. *Consider system (5.46) with (A2) and let $\alpha > 0$, $\gamma > 0$, $h_2 > h_1 \geq 0$ and $\mu \geq 1$ be given constants, if there exist positive definite and symmetric matrices $P_{i1} \in R^{n\times n}$, $P_{i2} \in R^{q\times q}$, $Z_{il} \in R^{m\times m}$, $Q_{il} \in R^{m\times m}$ and matrices $Y_{ik} \in R^{m\times m}$, $T_{ik} \in R^{m\times m}$, $L_{ik} \in R^{m\times m}$, $N_{i1} \in R^{q\times q}$, $N_{i2} \in R^{q\times r}$, $N_{i3} \in R^{s\times q}$, $N_{i4} \in R^{s\times r}$, $N_{i5} \in R^{s\times q}$, $N_{i6} \in R^{s\times r}$, $q < n, m = n+q$, $k = 1,2,3, l = 1,2, \forall i \in \mathscr{P}$ such that (5.51) is satisfied, (5.56) and the following inequalities hold,*

$$\Phi_i < \mu\Phi_j,\ Q_{il} \leq \mu Q_{jl},\ Z_{il} \leq \mu Z_{jl},\ \forall i,j \in \mathscr{P}. \tag{5.64}$$

$$\begin{bmatrix} \bar{\Omega}_i & \tilde{R}_i \\ * & -\tilde{W}_i \end{bmatrix} < 0, \tag{5.65}$$

where

$$\bar{\Omega}_i = \begin{bmatrix} \bar{\varphi}_{i11} & \tilde{\varphi}_{i12} & \tilde{\varphi}_{i13} \\ * & \hat{\varphi}_{i22} & \varphi_{i23} \\ * & * & -\gamma^2 I \end{bmatrix},$$

$$\bar{\varphi}_{i11} = \tilde{\varphi}_{i16} + \tilde{\varphi}_{i16}^T + \sum_{l=1}^{2} Q_{il} + Y_{i1} + Y_{i1}^T + \alpha\Phi_i,$$

and $\tilde{R}_i$, $\tilde{W}_i$, $\tilde{\varphi}_{i12}$, $\tilde{\varphi}_{i13}$ and $\tilde{\varphi}_{i16}$ are defined in Theorem 5.19. Then, there exists a q-order model (5.47) for an n-order original system (5.46) such that the error system (5.48) under ADT switching signals is exponentially stable with an H_∞ norm bound γ. Moreover, if LMIs (5.56), (5.64) and (5.65) have a feasible solution, then, switching signals with average dwell time τ_a are obtained by using (5.51) and an admissible q-order model (5.47) can be given by (5.58).

Proof. By using Lemma 5.18 and the approach similar to Theorem 5.19, the proof can be easily obtained and thus omitted.

Remark 5.23. In Theorems 5.19 and 5.22, the strict LMIs conditions are presented for exponential H_∞ model reduction problem for continuous-time switched systems with interval time-varying delay satisfying (A1) and (A2), respectively. When the information of the derivative of time delay is known, Lemma 5.17 is applicable. Otherwise, Lemma 5.18 can be applied. As is mentioned in [40], the cases (A1) and (A2) are more ubiquitous and practical than the case $0 \leq d(t) \leq d$ and $\dot{d}(t) \leq 1$, which is studied in [122]. From this point of view, the results in the present paper are more widely applicable than those in [122].

Remark 5.24. It is also pointed out that if there exists only one subsystem in such systems, that is, $p = 1$ denoting the systems with interval time-varying delay, our results are still applicable and less conservative than those in [133]. In addition, by making use of the proposed methods, it is possible to extend our results to general systems in the continuous-time domain such as linear systems, Markovian jump linear systems and so on. Hence, our methods are more flexible and less conservative.

5.3 Illustrative Examples

In this section, two examples are presented to check the validity of the proposed filtering design algorithms. Example 5.25 is for (A1) by means of Theorem 5.10 and Example 5.26 is for (A2) by using Corollary 5.12.

Example 5.25. Consider system (5.1) with (A1) consisting of two subsystems, where $h_2 = 0.6$, $h_1 = 0.2$, $d = 0.4$, subsystem 1 is described by

$$A_1 = \begin{bmatrix} -2 & 0 \\ 0 & -0.9 \end{bmatrix},$$
$$E_1 = \begin{bmatrix} -1 & 0 \\ 0 & -1 \end{bmatrix}, \; B_1 = \begin{bmatrix} 0 \\ 1 \end{bmatrix},$$
$$C_1 = \begin{bmatrix} 1 & 0 \end{bmatrix}, \; F_1 = \begin{bmatrix} 0 & 1 \end{bmatrix}, \; D_1 = 1,$$
$$H_1 = \begin{bmatrix} 1 & 2 \end{bmatrix}, \; G_1 = \begin{bmatrix} 1 & 0 \end{bmatrix}, \; M_1 = 0.1,$$

and subsystem 2 by

$$A_2 = \begin{bmatrix} -2 & 0 \\ 0 & -0.7 \end{bmatrix},$$
$$E_2 = \begin{bmatrix} -1 & 0 \\ -1 & -1 \end{bmatrix}, \; B_2 = \begin{bmatrix} -0.5 \\ 2 \end{bmatrix},$$
$$C_2 = \begin{bmatrix} 0 & 1 \end{bmatrix}, \; F_2 = \begin{bmatrix} 1 & 2 \end{bmatrix}, \; D_2 = 1,$$
$$H_2 = \begin{bmatrix} 2 & 1 \end{bmatrix}, \; G_2 = \begin{bmatrix} 0 & 1 \end{bmatrix}, \; M_2 = 0.2.$$

Our purpose is to design a filer and find out the admissible switching signals with average dwell time such that the filtering error system is exponentially stable with an H_∞ prescribed attenuation level γ. Hence, given $\alpha = 0.1$, $\mu = 1.5$ and $\gamma = 2.0$, by utilizing the LMI Toolbox, it follows from Theorem 5.10 that we can have a feasible solution. Furthermore, by (5.42), the desired filter can be obtained with the parameterized matrices as follows:

$$\left[\begin{array}{c|c} A_{1f} & B_{1f} \\ \hline C_{1f} & D_{1f} \end{array}\right] = \left[\begin{array}{cc|c} -7.6959 & 1.6883 & -0.5203 \\ 13.1591 & -11.0020 & 3.6454 \\ \hline 0.4740 & 1.8168 & -0.0462 \end{array}\right],$$

and

$$\left[\begin{array}{c|c} A_{2f} & B_{2f} \\ \hline C_{2f} & D_{2f} \end{array}\right] = \left[\begin{array}{cc|c} 4.5010 & 3.2723 & -0.9956 \\ -32.1685 & -15.9015 & 4.5911 \\ \hline 4.7236 & 0.7532 & 0.5802 \end{array}\right],$$

and the minimal average dwell time $\tau^* = 4.0547$ is given by (5.29). This means that a desired filter under the switching signals with average dwell time is developed by the proposed method. On the other hand, by solving Problem 5.15, the different γ^* and τ^* for the different μ and α can be obtained in Table 5.1 and Table 5.2, respectively. It is obviously seen from Table 5.1 that the exponential H_∞ performance index γ^* and the minimal average dwell time τ^* are dependent on μ when α is fixed. Moreover, it is also observed that the larger μ corresponds to the smaller γ^* while the larger τ^*. It is clearly seen from Table 5.2 that the exponential H_∞ performance index γ^* and the minimal average dwell time τ^* are dependent on α when μ is fixed. Moreover, it can be also observed that the larger α corresponds to the smaller τ^* while the larger γ^*.

Table 5.1 The different γ^* and τ^* for $\alpha = 0.1$

μ	1.2	1.3	1.4	1.5
τ^*	1.8232	2.6236	3.3647	4.0547
γ^*	1.8323	1.7677	1.7071	1.6527

Table 5.2 The different γ^* and τ^* for $\mu = 1.5$

α	0.15	0.2	0.25	0.3
τ^*	2.7031	2.0273	1.6219	1.3516
γ^*	1.6897	1.7281	1.7681	1.8095

Example 5.26. Consider system (5.1) with (A2) consisting of two subsystems, where $h_2 = 1.0$, $h_1 = 0.3$, subsystem 1 is given by

$$
\begin{aligned}
A_1 &= \begin{bmatrix} -0.2 & 0 \\ 0 & -0.8 \end{bmatrix}, \\
E_1 &= \begin{bmatrix} -1 & 0 \\ -1 & -1 \end{bmatrix}, \ B_1 = \begin{bmatrix} 0 \\ 1 \end{bmatrix}, \\
C_1 &= \begin{bmatrix} 1 & 0 \end{bmatrix}, \ F_1 = \begin{bmatrix} 0 & 1 \end{bmatrix}, \ D_1 = 1, \\
H_1 &= \begin{bmatrix} 1 & 0 \end{bmatrix}, \ G_1 = \begin{bmatrix} 1 & 0 \end{bmatrix}, \ M_1 = 0.1,
\end{aligned}
$$

and subsystem 2 by

$$
\begin{aligned}
A_2 &= \begin{bmatrix} -0.1 & 0 \\ 0 & -1 \end{bmatrix}, \\
E_2 &= \begin{bmatrix} -1 & 0 \\ 0 & -1 \end{bmatrix}, \ B_2 = \begin{bmatrix} -0.5 \\ 1 \end{bmatrix}, \\
C_2 &= \begin{bmatrix} 0 & 1 \end{bmatrix}, \ F_2 = \begin{bmatrix} 1 & 0 \end{bmatrix}, \ D_2 = 1, \\
H_2 &= \begin{bmatrix} 1 & 1 \end{bmatrix}, \ G_2 = \begin{bmatrix} 0 & 1 \end{bmatrix}, \ M_2 = 0.1.
\end{aligned}
$$

Given $\alpha = 0.25$, $\mu = 1.1$ and $\gamma = 8.0$, by Corollary 5.12, the desired filter matrices can be given by

$$
\left[\begin{array}{c|c} A_{1f} & B_{1f} \\ \hline C_{1f} & D_{1f} \end{array}\right] = \left[\begin{array}{cc|c} -3.7494 & -0.0671 & 0.2646 \\ 0.1767 & -0.8778 & -1.4389 \\ \hline 1.4795 & -0.1458 & 0.3370 \end{array}\right],
$$

and

$$
\left[\begin{array}{c|c} A_{2f} & B_{2f} \\ \hline C_{2f} & D_{2f} \end{array}\right] = \left[\begin{array}{cc|c} -0.8028 & 0.2889 & -0.7737 \\ -0.0974 & -3.4649 & 1.3508 \\ \hline 0.2278 & 2.1667 & -0.8187 \end{array}\right],
$$

and $\tau > \tau^* = 0.3812$. This implies that Corollary 5.12 is effective for system (5.1) with (A2). On the other hand, by solving Problem 5.16, Table 5.3 and Table 5.4

list the different γ^* and τ^* for different μ and α, respectively. It is obviously seen from Table 5.3 that the larger μ corresponds to the smaller γ^* while the larger τ^*. It is clearly observed from Table 5.4 that the larger α corresponds to the smaller τ^* while the larger γ^*.

Table 5.3 The different γ^* and τ^* for $\alpha = 0.25$

μ	1.2	1.3	1.4	1.5
τ^*	0.7293	1.0495	1.3459	1.6219
γ^*	5.7094	5.3394	5.0330	4.7722

Table 5.4 The different γ^* and τ^* for $\mu = 1.1$

α	0.1	0.15	0.2	0.25
τ^*	0.9531	0.6354	0.4766	0.3812
γ^*	4.0973	4.6398	5.3504	6.3289

Example 5.27. Consider system (5.46) with (A1) consisting of two subsystems, where $h_1 = 0.2$, $h_2 = 0.5$, $d = 1.1$, subsystem 1 is described by

$$A_1 = \begin{bmatrix} -0.5 & 0.2 & 0.3 \\ 0.1 & -0.8 & 0.4 \\ 0.2 & 0.3 & -0.9 \end{bmatrix},$$

$$E_1 = \begin{bmatrix} -0.7 & 0.6 & 0.2 \\ 0.1 & -0.2 & -0.4 \\ -0.3 & 0.2 & -0.6 \end{bmatrix},$$

$$B_1 = \begin{bmatrix} -0.3 & 0.1 & 0 \end{bmatrix}^T,$$
$$C_1 = \begin{bmatrix} 0.1 & 0.4 & -0.1 \end{bmatrix},$$
$$F_1 = \begin{bmatrix} 0.3 & -0.2 & 0.2 \end{bmatrix},$$
$$D_1 = 0.5,$$

and subsystem 2 by

$$A_2 = \begin{bmatrix} -0.9 & 0.1 & -0.3 \\ 0.4 & -0.2 & 0.2 \\ 0.1 & -0.2 & -0.3 \end{bmatrix},$$

$$E_2 = \begin{bmatrix} -0.5 & 0.2 & 0.1 \\ 0.2 & -0.9 & -0.3 \\ 0.1 & -0.5 & -0.9 \end{bmatrix},$$

$$B_2 = \begin{bmatrix} 0.1 & -0.3 & 0 \end{bmatrix}^T,$$
$$C_2 = \begin{bmatrix} 0.5 & 0.3 & -0.4 \end{bmatrix},$$
$$F_2 = \begin{bmatrix} 0.2 & -0.8 & 0.4 \end{bmatrix},$$
$$D_2 = 0.6.$$

Here, we are interested in designing a second-order system (5.47) and find out the admissible ADT switching signals such that the error system (5.48) is exponentially stable with an H_∞ norm bound γ. Hence, given $\alpha = 0.1$, $\mu = 1.2$ and $\gamma = 1.2$, by utilizing the LMI Toolbox, it follows from Theorem 5.19 that we can have a feasible solution. By (5.58), the desired reduced-order model can be obtained with the parameterized matrices as follows:

$$\left[\begin{array}{c|c|c} A_{r1} & E_{r1} & B_{r1} \\ \hline C_{r1} & F_{r1} & D_{r1} \end{array}\right] = \left[\begin{array}{cc|cc|c} -0.9245 & 0.1358 & -0.3470 & 0.4972 & 0.4929 \\ 0.0831 & -0.9322 & 0.0024 & -0.1566 & -0.0867 \\ \hline -0.0803 & -0.0594 & -0.1025 & -0.0519 & 0.4856 \end{array}\right],$$

and

$$\left[\begin{array}{c|c|c} A_{r2} & E_{r2} & B_{r2} \\ \hline C_{r2} & F_{r2} & D_{r2} \end{array}\right] = \left[\begin{array}{cc|cc|c} -0.9908 & 0.1400 & -0.4126 & 0.2448 & -0.1271 \\ 0.1248 & -0.9322 & 0.2333 & -0.3038 & 0.4831 \\ \hline -0.1575 & 0.1909 & -0.2178 & 0.1265 & 0.4728 \end{array}\right],$$

and the minimal ADT $\tau^* = 1.8232$ is given by (5.51). To illustrate the performances of the obtained reduced order models, when the control input is chosen as $u_i(t) = 20e^{-0.1t} sin(0.1t)$, switching signal ($\tau_a = 8.3957$) and output trajectories of the original system and the reduced-order model are shown in Figs. 5.1 and 5.2 with the zero initial conditions $\phi(\theta) = \begin{bmatrix} 0 & 0 & 0 \end{bmatrix}$, $\theta \in [-h_2, 0]$ and $\hat{x}(0) = \begin{bmatrix} 0 & 0 \end{bmatrix}$. This means that a desired reduced-order switched model under switching signals with ADT $\tau_a > \tau^*$ is developed by Theorem 5.19.

Example 5.28. Consider system (5.46) with (A2) consisting of two subsystems, where $h_2 = 0.7$, $h_1 = 0.1$, subsystem 1 is given by

$$A_1 = \begin{bmatrix} -0.9 & 0.4 & 0.2 \\ 0.2 & -0.7 & 0.3 \\ 0.4 & 0.1 & -0.8 \end{bmatrix},$$

$$E_1 = \begin{bmatrix} -0.6 & 0.3 & 0.1 \\ 0.1 & -0.5 & -0.2 \\ -0.1 & 0.3 & -1.2 \end{bmatrix}, \quad B_1 = \begin{bmatrix} -0.2 \\ 0.3 \\ 0 \end{bmatrix},$$

$$C_1 = \begin{bmatrix} -0.6 & 0 & 0 \end{bmatrix}, \quad F_1 = \begin{bmatrix} 1.0 & -1.4 & 0 \end{bmatrix}, \quad D_1 = 0.3,$$

and subsystem 2 by

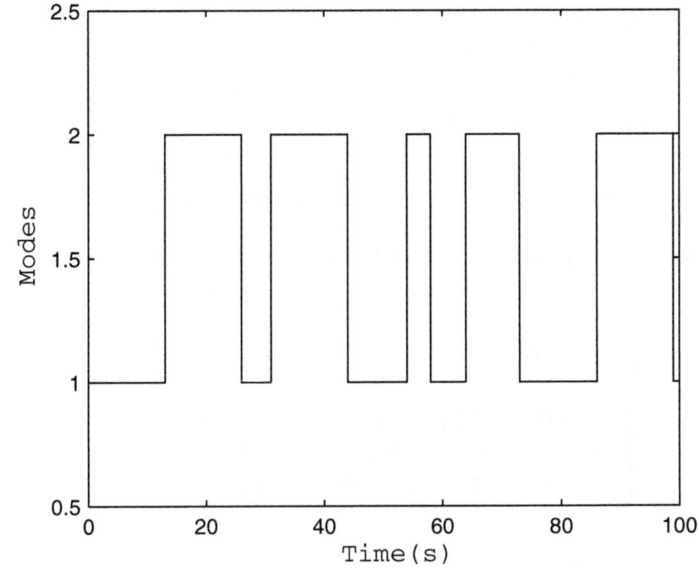

Fig. 5.1 Switching signal

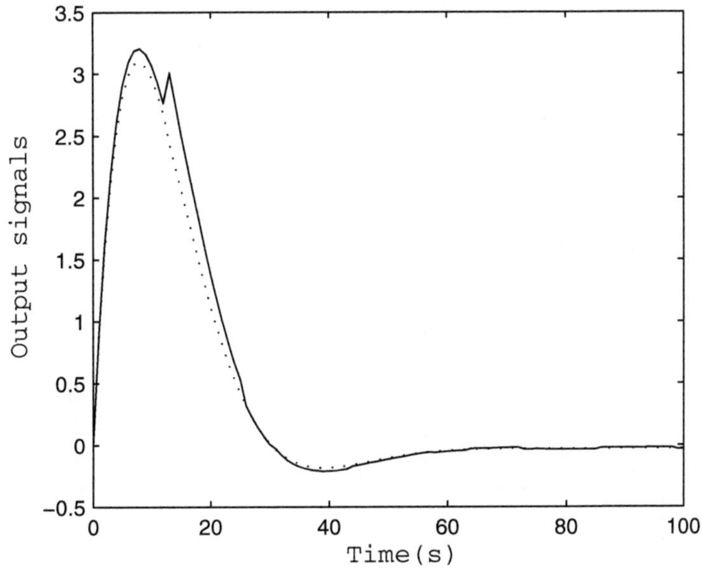

Fig. 5.2 $y(t)$(solid) and $\hat{y}(t)$(dash)

$$A_2 = \begin{bmatrix} -1.1 & 0.3 & -0.2 \\ -0.3 & -0.5 & 0.4 \\ 0.1 & -0.2 & -1.2 \end{bmatrix},$$
$$E_2 = \begin{bmatrix} -0.6 & 0.1 & -0.1 \\ -0.2 & -0.9 & 0.3 \\ 0.3 & -0.6 & -0.7 \end{bmatrix}, \ B_2 = \begin{bmatrix} 0.2 \\ -0.1 \\ 0 \end{bmatrix},$$
$$C_2 = \begin{bmatrix} 0 & -0.5 & 0.1 \end{bmatrix}, \ F_2 = \begin{bmatrix} 1.1 & 0 & -0.2 \end{bmatrix}, \ D_2 = 0.2.$$

Given $\alpha = 0.2$, $\mu = 1.8$ and $\gamma = 3.0$, by Theorem 5.22, the parameter matrices of a desired second-order model can be given by

$$\left[\begin{array}{c|c|c} A_{r1} & E_{r1} & B_{r1} \\ \hline C_{r1} & F_{r1} & D_{r1} \end{array}\right] = \left[\begin{array}{cc|cc|c} -0.9037 & 0.1761 & -0.4208 & 0.1634 & 0.2429 \\ 0.0995 & -0.8270 & 0.0770 & -0.2351 & -0.3606 \\ \hline 0.1050 & 0.1898 & -0.3017 & 0.7009 & 0.4841 \end{array}\right],$$

and

$$\left[\begin{array}{c|c|c} A_{r2} & E_{r2} & B_{r2} \\ \hline C_{r2} & F_{r2} & D_{r2} \end{array}\right] = \left[\begin{array}{cc|cc|c} -1.0790 & -0.0266 & -0.6870 & 0.2313 & -0.3158 \\ -0.1810 & -0.4432 & -0.4836 & -0.8996 & 0.4084 \\ \hline -0.4043 & 0.2531 & -1.1632 & 0.1230 & 0.1283 \end{array}\right],$$

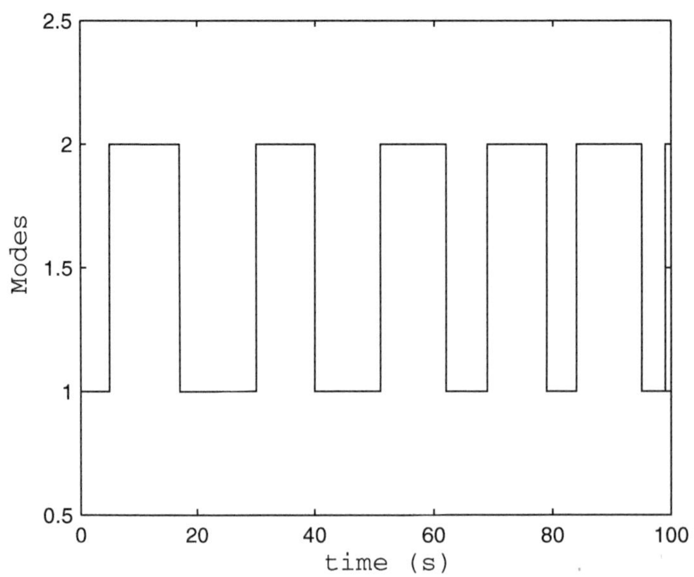

Fig. 5.3 Switching signal

and the minimal ADT $\tau^* = 2.9389$. To demonstrate the performances of the obtained reduced order models, under the control input as $u_i(t) = 10e^{-0.1t}sin(0.5t)$ and the zero initial conditions, switching signal ($\tau_a = 8.3846$) and output trajectories of the original system and the reduced-order model are shown in Figs. 5.3 and 5.4. This implies that Theorem 5.22 is effective for system (5.46) with (A2).

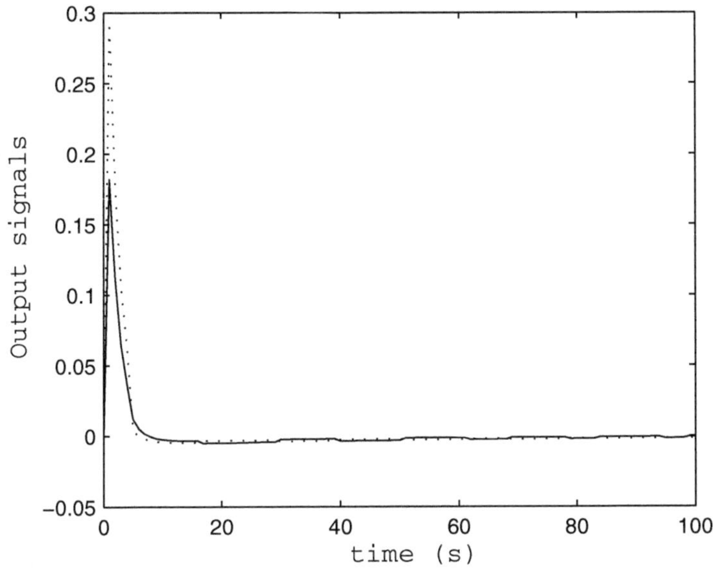

Fig. 5.4 $y(t)$(solid) and $\hat{y}(t)$(dash)

From the simulation results above, it can be seen that the obtained reduced-order model can approximate the original model very well. Therefore, the proposed method is effective.

5.4 Summary

We have studied the problem of the exponential H_∞ filter design for switched systems with interval time-varying delay. First, we have considered the case where the time-varying delay is differentiable and its derivative is bounded. Based on average dwell time method and the free-weighting matrix approach, sufficient conditions for the existence of a filter assuring the resultant filtering error system to be exponentially stable with an H_∞ prescribed attenuation level are obtained and formulated in terms of LMIs. The desired filter has been constructed by solving the corresponding LMIs. Secondly, the obtained criteria have also been extended to the case where the time-varying delay is continuous and bounded, that is, there is no restriction on the derivative of interval time-varying delay. Simulation results have been provided to illustrate the effectiveness of the proposed methods.

Part II

Fault Detection

6

Robust Fault Detection for Discrete-Time Switched Linear Systems with State Delays

This chapter investigates the problem of robust fault detection for discrete-time switched systems with state delays under arbitrary switching signal. The fault detection filter is employed as the residual generator whose parameters are dependent on the system mode. The objective of this chapter is to design the robust fault detection filter in order to minimize the estimation error between the residuals and the faults for unknown inputs, control inputs and model uncertainties. By switched Lyapunov functional approach, a sufficient condition for the solvability of this problem is established.

6.1 Introduction

Fault Detection and Isolation (FDI) in dynamic systems has attracted great attention of many researchers over the past decades, and some model-based fault detection approaches have been proposed (see, e.g. [87, 7, 8, 23] and the references therein). Among these model-based approaches, the most common one is to use state observers or filters to construct a residual signal and compare it with a predefined threshold. When the residual evaluation function has a value larger than the threshold, an alarm is generated. It is well known that unknown inputs, control inputs, model uncertainties and the faults are coupled in many industrial systems, which is a source of false alarms. This means that FDI systems have to be sensitive to faults and simultaneously robust to unknown inputs, control inputs and model uncertainties. Therefore, it is of great significance to design a robust FDI system [80, 12]. Recently, an H_∞-filtering formulation of FDI problem has been presented to solve the robust FDI problem [151, 152, 39]. In [152], the problem of Robust Fault Detection Filter Design (RFDFD) for discrete-time Markovain jump linear systems is formulated as an H_∞-filtering problem. In [39], the problem of RFDFD for discrete-time networked systems with multiple state delays and unknown input is transformed into an H_∞-filtering problem for Markovain jumping system. In this paper, the problem of RFDFD for discrete-time switched systems with state delays is cast into an H_∞ filtering problem. However, to the best of the authors' knowledge,

D. Wang et al.: *Robust Filtering & Fault Detect. of Switched Delay Systems*, LNCIS 445, pp. 107–120.
DOI: 10.1007/978-3-642-37685-6_6

the problem of RFDFD for discrete-time switched systems with state delays has not been investigated yet. This motivates us to study this interesting and challenging problem, which has great potential in practical applications.

This chapter studies the problem of RFDFD by using switched Lyapunov functional approach for discrete-time switched systems with state delays. Firstly, the residual generator is constructed based on the filter. Secondly, by augmenting the states of the original system and the fault detection filter, the problem of RFDFD is formulated as an H_∞ filtering problem. The objective is to make the difference between the faults and the residuals as small as possible, and increase robustness of the residuals to the unknown input. Then, by using switched Lyapunov functional approach, a sufficient condition on the existence of such filters is established and the desired filters are constructed. Finally, a simulation example is presented to demonstrate the effectiveness of the proposed method.

The rest of this chapter is organized as follows. In Section 6.2, system descriptions and definitions are presented. A sufficient condition on the existence of a robust fault detection filter for discrete-time switched systems with state delays is presented in terms of LMIs, and the desired filter is constructed in Section 6.3. To demonstrate the validity of the proposed approach, an example is given in Section 6.4 which is followed by a conclusion in Section 6.5.

6.2 Problem Formulation

Consider the following discrete-time switched systems with state delays:

$$\begin{aligned} x_{k+1} &= \sum_{i=1}^{N} \xi_i(k)\left(A_i x_k + A_{di} x_{k-d} + E_i u_k + B_i d_k + G_i f_k\right), \\ y_k &= \sum_{i=1}^{N} \xi_i(k)\left(C_i x_k + C_{di} x_{k-d} + Q_i u_k + D_i d_k + J_i f_k\right), \end{aligned} \tag{6.1}$$

where $x_k \in R^n$ is the state, $y_k \in R^r$ is the measured output, $d_k \in R^p$, $u_k \in R^s$ and $f_k \in R^q$ are, respectively, the unknown input, the control input and the faults which belong to $l_2[0,\infty)$. The matrices A_i, A_{di}, B_i, C_i, C_{di}, D_i, E_i, G_i, J_i and Q_i are of appropriate dimensions, where $A_i = \bar{A}_i + \Delta A_i(k), A_{di} = \bar{A}_{di} + \Delta A_{di}(k)$ and $E_i = \bar{E}_i + \Delta E_i(k)$. The modeling errors $\Delta A_i(k), \Delta A_{di}(k)$ and $\Delta E_i(k)$ are norm-bounded uncertainties satisfying

$$\left[\Delta A_i(k)\ \Delta A_{di}(k)\ \Delta E_i(k)\right] = \bar{H}_i \bar{F}(k)\left[\bar{C}_{1i}\ \bar{C}_{2i}\ \bar{C}_{3i}\right],$$

where $\bar{H}_i, \bar{C}_{1i}, \bar{C}_{2i}, \bar{C}_{3i}$ are known constant matrices, while $\bar{F}(k)$ is an unknown time-varying function matrix satisfying $\bar{F}^T(k)\bar{F}(k) \leq I$. The positive integers N and d denote the number of subsystems and the state delay, respectively. $\xi_i(k) : Z^+ \to \{0,1\}$ where

$$\sum_{i=1}^{N} \xi_i(k) = 1, k \in Z^+, i \in \mathscr{N} = \{1, 2, \cdots, N\},$$

is the switching signal which specifies which subsystem is activated at the switching instant.

An FDI system consists of a residual generator and a residual evaluation stage including an evaluation function and a threshold. For the purpose of residual generation, the following fault detection filter is constructed as a residual generator:

$$\begin{aligned} \hat{x}_{k+1} &= \sum_{i=1}^{N} \xi_i(k)\left(A_{fi}\hat{x}_k + B_{fi}y_k\right), \\ r_k &= \sum_{i=1}^{N} \xi_i(k)\left(C_{fi}\hat{x}_k + D_{fi}y_k\right), \end{aligned} \tag{6.2}$$

where $\hat{x}_k$ is the filter's state, r_k is the residual signal. The matrices A_{fi}, B_{fi}, C_{fi} and D_{fi} are the filter parameters to be determined.

For the purpose of fault detection, it is not necessary to estimate the fault f_k. Sometimes one is more interested in the fault signal of a certain frequency interval, which can be formulated as the weighted fault $\hat{f}(z) = W_f(z)f(z)$ with $W_f(z)$ being a given stable weighting matrix. A minimal realization of $\hat{f}(z) = W_f(z)f(z)$ is supposed to be

$$\begin{aligned} \bar{x}_{k+1} &= A_w\bar{x}_k + B_wf_k, \\ \hat{f}_k &= C_w\bar{x}_k + D_wf_k, \end{aligned} \tag{6.3}$$

where $\bar{x}_k \in \mathscr{R}^{\bar{n}}$ is the state of the weighted fault, $f_k \in R^q$ is the original fault and $\hat{f}_k \in R^q$ is the weighted fault. A_w, B_w, C_w and D_w are known constant matrices.

Remark 6.1. The parameters of filter (6.2) depend on the system modes. This means that the switching signals in system (6.1) and filter (6.2) are the same. In practice, each subsystem of filter (6.2) is designed for the corresponding subsystem of system (6.1). Then, each pair of subsystems can be activated by one signal. As in [15, 10], it is assumed that the switching signal is not known a priori, but its instantaneous value is available in real-time implementation. When the switching signal changes, the jumps between the modes occur simultaneously and in pairs.

Remark 6.2. Similar to [152] and [39], the introduction of $W_f(z)$ could limit the frequency ranges of interest, but the system performance could be improved and the frequency characteristics required to reflect the emphases of different frequency ranges could be captured.

Denoting $e_k = r_k - \hat{f}_k$ and augmenting the model of system (6.1) to include the states of (6.2), we can obtain the augmented system as follows:

$$\begin{aligned} \tilde{x}_{k+1} &= \sum_{i=1}^{N} \xi_i(k)\left(\tilde{A}_i\tilde{x}_k + \tilde{A}_{di}\tilde{x}_{k-d} + \tilde{B}_iw_k\right), \\ e_k &= \sum_{i=1}^{N} \xi_i(k)\left(\tilde{C}_i\tilde{x}_k + \tilde{C}_{di}\tilde{x}_{k-d} + \tilde{D}_iw_k\right), \end{aligned} \tag{6.4}$$

where

$$\tilde{x}_k = \begin{bmatrix} x_k \\ \hat{x}_k \\ \bar{x}_k \end{bmatrix}, \; w_k = \begin{bmatrix} u_k \\ d_k \\ f_k \end{bmatrix},$$
$$\tilde{A}_i = \begin{bmatrix} A_i & 0 & 0 \\ B_{fi}C_i & A_{fi} & 0 \\ 0 & 0 & A_w \end{bmatrix},$$
$$\tilde{A}_{di} = \begin{bmatrix} A_{di} & 0\,0 \\ B_{fi}C_{di} & 0\,0 \\ 0 & 0\,0 \end{bmatrix},$$
$$\tilde{B}_i = \begin{bmatrix} E_i & B_i & G_i \\ B_{fi}Q_i & B_{fi}D_i & B_{fi}J_i \\ 0 & 0 & B_w \end{bmatrix},$$
$$\tilde{C}_i = \begin{bmatrix} D_{fi}C_i & C_{fi} & -C_w \end{bmatrix},$$
$$\tilde{C}_{di} = \begin{bmatrix} D_{fi}C_{di} & 0 & 0 \end{bmatrix},$$
$$\tilde{D}_i = \begin{bmatrix} D_{fi}Q_i & D_{fi}D_i & D_{fi}J_i - D_w \end{bmatrix}.$$

Remark 6.3. It should be noted from $e_k = r_k - \hat{f}_k$ that the residual r_k generated by filter (6.2) provides an estimate of the fault $\hat{f}_k$. The stable weighting matrix $W_f(z)$ is given. Thus, detection and isolation of the fault f_k can be achieved by examining the values of the residual r_k. That is, the designed filter not only detects the occurred fault, but also can isolate it.

Now, the problem of RFDFD can be transformed into an H_∞ filtering problem for system (6.4): to develop filter (6.2) for system (6.1) such that the augmented system (6.4) under arbitrary switching signal is asymptotically stable when $w_k = 0$ and, under zero-initial condition, the infimum of γ is made small in the feasibility of

$$\sup_{w_k \neq 0, w_k \in l_2[0,\infty)} \frac{\|e_k\|_2}{\|w_k\|_2} < \gamma, \gamma > 0 \tag{6.5}$$

After designing the residual generator, the remaining important task is to evaluate the generated residual. One of the widely adopted approaches is to select a threshold and a residual evaluation function. In this paper, the residual evaluation function is chosen as

$$J_L(r) = \|r_k\|_{2,L} = \left(\sum_{k=l_0}^{l_0+L} r_k^T r_k \right)^{1/2}, \tag{6.6}$$

where l_0 is the initial evaluation time instant, L is the evaluation time window.

Remark 6.4. In fact, the length of the evaluation window L is limited since it is desired that the faults will be detected as early as possible, while an evaluation of residual signal over the whole time range is not practical. This point has been mentioned in [12, 151, 152].

Once the evaluation function has been selected, we are able to determine the threshold. Since the faults can be detected by using the following logical relationship:

$$J_L(r) > J_{th} \Rightarrow Faults \Rightarrow Alarm,$$
$$J_L(r) \le J_{th} \Rightarrow No\ Fault,$$

it is reasonable to choose the threshold as

$$J_{th} = \sup_{d \in l_2,\, u \in l_2,\, f=0} \|r_k\|_{2,L}. \tag{6.7}$$

It is clear that the computation of J_{th} involves the determination of the unknown inputs d_k and the control inputs u_k on the residuals r_k.

6.3 Fault Detection Filter Design

In this section, a sufficient condition on the existence of the robust fault detection filters would be given and a desired filer could be constructed.

Lemma 6.5. *[96]: Let Y, H and C be matrices with appropriate dimensions. Suppose that Y is symmetric and $\Delta(k)^T\Delta(k) \le I$, then,*

$$Y + H\Delta(k)C + C^T\Delta(k)H < 0,$$

if and only if

$$Y + \varepsilon^{-1}HH^T + \varepsilon C^T C < 0,$$

where ε is a given positive scalar.

Lemma 6.6. *For a given scalar $\gamma > 0$, system (6.4) under arbitrary switching signals is asymptotically stable when $w_k = 0$ and, under zero-initial conditions, guarantees the performance index (6.5) for all nonzero $w_k \in l_2[0,\infty)$, if there exist symmetric and positive definite matrices P_i and Q_i, $i \in \mathscr{N}$ such that the following inequality holds,*

$$\begin{bmatrix} -P_j^{-1} & \tilde{A}_i & \tilde{A}_{di} & \tilde{B}_i & 0 & 0 \\ * & -P_i & 0 & 0 & \tilde{C}_i^T & I \\ * & * & -Q_l & 0 & \tilde{C}_{di}^T & 0 \\ * & * & * & -\gamma^2 I & \tilde{D}_i^T & 0 \\ * & * & * & * & -I & 0 \\ * & * & * & * & * & -Q_i^{-1} \end{bmatrix} < 0, \tag{6.8}$$

Proof. First, the asymptotical stability of system (6.4) with $w_k = 0$ is established. The following Lyapunov functional is constructed by:

$$V_k = \tilde{x}_k^T \left(\sum_{i=1}^{N} \xi_i(k) P_i \right) \tilde{x}_k + \sum_{s=k-d}^{k-1} \tilde{x}_s^T \left(\sum_{i=1}^{N} \xi_i(k) Q_i \right) \tilde{x}_s, \tag{6.9}$$

where P_i and Q_i are symmetric and positive definite matrices.

Then, we get

$$\begin{aligned}
\Delta V_k &= V_{k+1} - V_k \\
&= \tilde{x}_{k+1}^T \left(\sum_{i=1}^{N} \xi_i (k+1) P_i \right) \tilde{x}_{k+1} \\
&\quad - \tilde{x}_k^T \left(\sum_{i=1}^{N} \xi_i (k) P_i \right) \tilde{x}_k + \tilde{x}_k^T \left(\sum_{i=1}^{N} \xi_i (k) Q_i \right) \tilde{x}_k \\
&\quad - \tilde{x}_{k-d}^T \left(\sum_{i=1}^{N} \xi_i (k-d) Q_i \right) \tilde{x}_{k-d}.
\end{aligned} \tag{6.10}$$

As this has to be satisfied under arbitrary switching signals, it follows that this has to hold for the configuration

$$\begin{aligned}
&\xi_i (k) = 1,\ \xi_{r \neq i} (k) = 0,\ \xi_j (k+1) = 1, \\
&\xi_{r \neq j} (k+1) = 0,\ \xi_l (k-d) = 1,\ \xi_{r \neq l} (k-d) = 0.
\end{aligned}$$

Along the solution of system (6.4) with $w_k = 0$, we have

$$\Delta V_k|_{w_k=0} = \tilde{\eta}_k^T \Lambda_i \tilde{\eta}_k, \tag{6.11}$$

where

$$\begin{aligned}
\tilde{\eta}_k^T &= \left[\tilde{x}_k^T\ \tilde{x}_{k-d}^T \right], \\
\Lambda_i &= \begin{bmatrix} \tilde{A}_i^T \\ \tilde{A}_{di}^T \end{bmatrix} P_j \left[\tilde{A}_i\ \tilde{A}_{di} \right] + \begin{bmatrix} Q_i - P_i & 0 \\ * & -Q_l \end{bmatrix}.
\end{aligned}$$

By using Schur complement, it follows from (6.8) that $\Lambda_i < 0$. Thus, from (6.11) we have $\Delta V_k|_{w_k=0} < -\rho \left\| x_k \right\|^2$ for a sufficiently small $\rho > 0$ and $x_k \neq 0$, which establishes the asymptotical stability of system (6.4).

Secondly, we consider the following performance index:

$$J = \sum_{k=0}^{K-1} \left[e_k^T e_k - \gamma^2 w_k^T w_k \right], \tag{6.12}$$

where K is an arbitrary positive integer.

For any nonzero $w_k \in l_2 [0, \infty)$ and under zero-initial condition $\tilde{x}_0 = 0$, one has

$$\begin{aligned}
J &= \sum_{k=0}^{K-1} \left[e_k^T e_k - \gamma^2 w_k^T w_k + \Delta V_k|_{(6.4)} \right] - V_K \\
&\leq \sum_{k=0}^{K-1} \left[e_k^T e_k - \gamma^2 w_k^T w_k + \Delta V_k|_{(6.4)} \right],
\end{aligned} \tag{6.13}$$

where $\Delta V_k|_{(6.4)}$ defines ΔV_k along the solution of system (6.4).

It is noted that

$$e_k^T e_k - \gamma^2 w_k^T w_k + \Delta V_k|_{(6.4)} = \eta_k^T \Pi \eta_k, \tag{6.14}$$

where

$$\begin{aligned}
\Pi = & \begin{bmatrix} \tilde{A}_i^T \\ \tilde{A}_{di}^T \\ \tilde{B}_i^T \end{bmatrix} P_j \left[\tilde{A}_i \; \tilde{A}_{di} \; \tilde{B}_i \right] \\
& + \begin{bmatrix} -P_i + Q_i & 0 & 0 \\ * & -Q_l & 0 \\ * & * & -\gamma^2 I \end{bmatrix} \\
& + \begin{bmatrix} \tilde{C}_i^T \\ \tilde{C}_{di}^T \\ \tilde{D}_i^T \end{bmatrix} \left[\tilde{C}_i \; \tilde{C}_{di} \; \tilde{D}_i \right], \\
\eta_k = & \left[\tilde{x}_k^T \; \tilde{x}_{k-d}^T \; w_k^T \right]^T .
\end{aligned}$$

It follows from (6.8) and Schur complement that $\Pi < 0$, which implies $J < 0$. Then, one has that for any nonzero $w_k \in l_2[0, \infty)$, $\|e_k\|_2 < \gamma \|w_k\|_2$, which completes the proof.

Lemma 6.7. *For a given scalar $\gamma > 0$ and the augmented system (6.4), LMI (6.8) is feasible, if there exist symmetric and positive definite matrices R_i, Φ_i, and matrices Ω_i, $i \in \mathcal{N}$ such that the following inequality holds,*

$$\begin{bmatrix} -R_j & \tilde{A}_i \Omega_i & \tilde{A}_{di} \Omega_i & \tilde{B}_i & 0 & 0 \\ * & \bar{R}_i & 0 & 0 & \Omega_i^T \tilde{C}_i^T & \Omega_i^T \\ * & * & \bar{\Phi}_l & 0 & \Omega_i^T \tilde{C}_{di}^T & 0 \\ * & * & * & -\gamma^2 I & \tilde{D}_i^T & 0 \\ * & * & * & * & -I & 0 \\ * & * & * & * & * & -\Phi_i \end{bmatrix} < 0, \tag{6.15}$$

where

$$\begin{aligned}
\bar{R}_i &= R_i - \left(\Omega_i^T + \Omega_i \right), \\
\bar{\Phi}_l &= \Phi_l - \left(\Omega_i^T + \Omega_i \right).
\end{aligned}$$

Proof. Assume that (6.15) is feasible, then it is easy to see that

$$R_i - \left(\Omega_i^T + \Omega_i \right) < 0,$$

which means that Ω_i is nonsingular. Since $R_i > 0$, we have

$$(R_i - \Omega_i)^T R_i^{-1} (R_i - \Omega_i) \geq 0,$$

which implies

$$-\Omega_i^T R_i^{-1} \Omega_i \leq R_i - \left(\Omega_i + \Omega_i^T\right).$$

Similarly, one has

$$-\Omega_i^T \Phi_l^{-1} \Omega_i \leq \Phi_l - \left(\Omega_i + \Omega_i^T\right).$$

Then, (6.15) is transformed into

$$\begin{bmatrix} -R_j & \tilde{A}_i\Omega_i & \tilde{A}_{di}\Omega_i & \tilde{B}_i & 0 & 0 \\ * & \hat{R}_i^{-1} & 0 & 0 & \Omega_i^T \tilde{C}_i^T & \Omega_i^T \\ * & * & \hat{\Phi}_l^{-1} & 0 & \Omega_i^T \tilde{C}_{di}^T & 0 \\ * & * & * & -\gamma^2 I & \tilde{D}_i^T & 0 \\ * & * & * & * & -I & 0 \\ * & * & * & * & * & -\Phi_i \end{bmatrix} < 0, \tag{6.16}$$

where

$$\hat{R}_i^{-1} = -\Omega_i^T R_i^{-1} \Omega_i,$$
$$\hat{\Phi}_l^{-1} = -\Omega_i^T \Phi_l^{-1} \Omega_i.$$

By means of premultiplying

$$diag\left\{I,\, \Omega_i^{-T},\, \Omega_i^{-T},\, I,\, I,\, I\right\},$$

and postmultiplying

$$diag\left\{I, \Omega_i^{-1},\, \Omega_i^{-1},\, I, I, I\right\},$$

to (6.16) yields

$$\begin{bmatrix} -R_j & \tilde{A}_i & \tilde{A}_{di} & \tilde{B}_i & 0 & 0 \\ * & -R_i^{-1} & 0 & 0 & \tilde{C}_i^T & I \\ * & * & -\Phi_l^{-1} & 0 & \tilde{C}_{di}^T & 0 \\ * & * & * & -\gamma^2 I & \tilde{D}_i^T & 0 \\ * & * & * & * & -I & 0 \\ * & * & * & * & * & -\Phi_i \end{bmatrix} < 0. \tag{6.17}$$

Letting $R_i = P_i^{-1}$ and $\Phi_i = Q_i^{-1}$, by means of Schur Complement, we can see that (6.17) is equivalent to (6.8). The proof is completed.

Remark 6.8. We can see from (6.8) in Lemma 6.6 that it is difficult to deal with the problem of RFDFD due to the existence of P_j^{-1}, which leads to some product terms between P_j and $\tilde{A}_i$, $\tilde{A}_{di}$ and $\tilde{B}_i$. To overcome the difficulties, an auxiliary slack matrix Ω_i is introduced in Lemma 6.7 such that these product terms are decoupled. That is, R_i and R_j are not involved in any product with $\tilde{A}_i$, $\tilde{A}_{di}$ and $\tilde{B}_i$ in (6.15). This implies it is more tractable to cope with the problem of RFDFD.

Now, we will present our main results in this paper as follows.

Theorem 6.9. *For a given scalar $\gamma > 0$, the problem of RFDFD for system (6.1) is solvable, if there exist symmetric and positive definite matrices R_{1i}, R_{4i}, R_{6i}, X_{1i}, X_{4i}, X_{6i}, and metrics R_{2i}, R_{3i}, R_{5i}, X_{2i}, X_{3i}, X_{5i}, Z_i, Y_i, W_i, H_i, M_i, L_i, N_i, S_i and a scalar $\varepsilon > 0$, $i \in \mathscr{N}$ such that the following LMI holds,*

$$\begin{bmatrix} -\Psi_{11} & \Psi_{12} & \Psi_{13} & \Psi_{14} & 0 & 0 & \Psi_{17} \\ * & \hat{\Psi}_{22} & \Psi_{23} & \Psi_{24} & \Psi_{25}^T & \Psi_{26}^T & 0 \\ * & * & \hat{\Psi}_{33} & \Psi_{34} & \Psi_{35}^T & 0 & 0 \\ * & * & * & \hat{\Psi}_{44} & \Psi_{45}^T & 0 & 0 \\ * & * & * & * & -I & 0 & 0 \\ * & * & * & * & * & -\Psi_{66} & 0 \\ * & * & * & * & * & * & -\varepsilon I \end{bmatrix} < 0, \tag{6.18}$$

where

$$\Psi_{11} = \begin{bmatrix} R_{1j} & R_{2j} & R_{3j} \\ * & R_{4j} & R_{5j} \\ * & * & R_{6j} \end{bmatrix},$$

$$\Psi_{12} = \begin{bmatrix} Z_i\bar{A}_i & Z_i\bar{A}_i & 0 \\ \hat{Y}_i + L_i & \hat{Y}_i & 0 \\ 0 & 0 & W_iA_w \end{bmatrix},$$

$$\Psi_{13} = \begin{bmatrix} Z_i\bar{A}_{di} & Z_i\bar{A}_{di} & 0 \\ \hat{Y}_{di} & \hat{Y}_{di} & 0 \\ 0 & 0 & 0 \end{bmatrix},$$

$$\Psi_{14} = \begin{bmatrix} Z_i\bar{E}_i & Z_iB_i & Z_iG_i \\ Y_{ei} & Y_{bi} & Y_{gi} \\ 0 & 0 & W_iB_w \end{bmatrix},$$

$$\Psi_{17}^T = \begin{bmatrix} \bar{H}_i^T Z_i^T & \bar{H}_i^T Y_i^T & 0 \end{bmatrix}^T,$$

$$\hat{\Psi}_{22} = \Psi_{22} - \Psi_{26} - \Psi_{26}^T,$$

$$\Psi_{22} = \begin{bmatrix} R_{1i} + \varepsilon\bar{C}_{1i}^T\bar{C}_{1i} & R_{2i} + \varepsilon\bar{C}_{1i}^T\bar{C}_{1i} & R_{3i} \\ * & R_{4i} + \varepsilon\bar{C}_{1i}^T\bar{C}_{1i} & R_{5i} \\ * & * & R_{6i} \end{bmatrix},$$

$$\Psi_{26} = \begin{bmatrix} Z_i & Z_i & 0 \\ Y_i + M_i & Y_i & 0 \\ 0 & 0 & W_i \end{bmatrix},$$

$$\Psi_{23} = \begin{bmatrix} \bar{C}_{12i} & \bar{C}_{12i} & 0 \\ \bar{C}_{12i} & \bar{C}_{12i} & 0 \\ 0 & 0 & 0 \end{bmatrix},$$

$$\Psi_{24} = \begin{bmatrix} \bar{C}_{13i} & 0 & 0 \\ \bar{C}_{13i} & 0 & 0 \\ 0 & 0 & 0 \end{bmatrix},$$

$$\Psi_{25} = \begin{bmatrix} N_iC_i + S_i & N_iC_i & -C_w \end{bmatrix},$$

$$
\begin{aligned}
\hat{\Psi}_{33} &= \Psi_{33} - \Psi_{26} - \Psi_{26}^T,\\
\Psi_{33} &= \begin{bmatrix} X_{1l} + \varepsilon \bar{C}_{2i}^T \bar{C}_{2i} & X_{2l} + \varepsilon \bar{C}_{2i}^T \bar{C}_{2i} & X_{3l} \\ * & X_{4l} + \varepsilon \bar{C}_{2i}^T \bar{C}_{2i} & X_{5l} \\ * & * & X_{6l} \end{bmatrix},\\
\Psi_{34} &= \begin{bmatrix} \bar{C}_{23i} & 0 & 0 \\ \bar{C}_{23i} & 0 & 0 \\ 0 & 0 & 0 \end{bmatrix},\\
\Psi_{35} &= \begin{bmatrix} N_i C_{di} & N_i C_{di} & 0 \end{bmatrix},\\
\hat{\Psi}_{44} &= -\gamma^2 I + \varepsilon C_{3i}^T C_{3i},\\
\Psi_{45} &= \begin{bmatrix} N_i Q_i & N_i D_i & N_i J_i - D_w \end{bmatrix},\\
\Psi_{66} &= \begin{bmatrix} X_{1i} & X_{2i} & X_{3i} \\ * & X_{4i} & X_{5i} \\ * & * & X_{6i} \end{bmatrix},\\
\hat{Y}_i &= Y_i \bar{A}_i + H_i C_i,\\
\hat{Y}_{di} &= Y_i \bar{A}_{di} + H_i C_{di},\\
Y_{ei} &= Y_i \bar{E}_i + H_i Q_i,\\
Y_{bi} &= Y_i B_i + H_i D_i,\\
Y_{gi} &= Y_i G_i + H_i J_i,\\
\bar{C}_{12i} &= \varepsilon \bar{C}_{1i}^T \bar{C}_{2i},\\
\bar{C}_{13i} &= \varepsilon \bar{C}_{1i}^T \bar{C}_{3i},\\
\bar{C}_{23i} &= \varepsilon \bar{C}_{2i}^T \bar{C}_{3i},
\end{aligned}
$$

then, a robust fault detection filter (6.2) can be constructed by

$$
\begin{bmatrix} A_{fi} & B_{fi} \\ C_{fi} & D_{fi} \end{bmatrix} = \begin{bmatrix} V_i^{-1} L_i M_i^{-1} V_i & V_i^{-1} H_i \\ S_i M_i^{-1} V_i & N_i \end{bmatrix}, \tag{6.19}
$$

where $V_i \in \mathscr{R}^{n \times n}$ is any invertible matrix (for example, V_i could be set as I).

Proof. By Lemmas 6.6 and 6.7, the augmented system (6.4) under arbitrary switching signals is asymptotically stable when $w_k = 0$ and, under zero-initial conditions, guarantees (6.5) for all nonzero $w_k \in l_2[0,\infty)$, if LMI (6.15) holds.

Note that from (6.18), we have

$$
\begin{bmatrix} Z_i + Z_i^T & Z_i + Y_i^T + M_i^T & 0 \\ * & Y_i + Y_i^T & 0 \\ * & * & W_i + W_i^T \end{bmatrix} > 0, \tag{6.20}
$$

which means that Z_i, Y_i and W_i are nonsingular.

Premultiplying $\begin{bmatrix} I & -I & 0 \end{bmatrix}$ and postmultiplying $\begin{bmatrix} I & -I & 0 \end{bmatrix}^T$ to (6.20), one has $-M_i - M_i^T > 0$, which implies that M_i is nonsingular. And thus, if (6.18) holds, there exist nonsingular matrices V_i and U_i satisfying $M_i = V_i U_i$.

Now, introduce

$$F_i^T = \begin{bmatrix} Z_i & 0 & 0 \\ Y_i & V_i & 0 \\ 0 & 0 & W_i \end{bmatrix}, \ \Omega_i = \begin{bmatrix} I & I & 0 \\ U_i & 0 & 0 \\ 0 & 0 & I \end{bmatrix} F_i^{-1},$$

and define

$$\begin{aligned} &H_i \triangleq V_i B_{fi}, \ L_i \triangleq V_i A_{fi} U_i, \\ &S_i \triangleq C_{fi} U_i, \ N_i \triangleq D_{fi}, \ M_i \triangleq V_i U_i, \\ &R_j \triangleq F_i^{-T} \Psi_{11} F_i^{-1}, \ R_i \triangleq F_i^{-T} \Psi_{22} F_i^{-1}, \\ &\Phi_l \triangleq F_i^{-T} \Psi_{33} F_i^{-1}, \ \Phi_i \triangleq F_i^{-T} \Psi_{66} F_i^{-1}. \end{aligned} \tag{6.21}$$

By using (6.4), (6.18) and (6.21), we can get

$$\begin{aligned} &F_i^T \tilde{A}_i \Omega_i F_i = \Psi_{12}, \ F_i^T \tilde{A}_{di} \Omega_i F_i = \Psi_{13}, \\ &F_i^T \tilde{B}_i = \Psi_{14}, \ F_i^T \Omega_i F_i = \Psi_{26}, \\ &\tilde{C}_i \Omega_i F_1 = \Psi_{25}, \ \tilde{C}_{di} \Omega_i F_i = \Psi_{35}. \end{aligned} \tag{6.22}$$

Then, by using (6.22), Lemma 6.5 and Schur complement, performing a congruence transformation to (6.15) via $diag\{F_i, F_i, F_i, I, I, F_i\}$ yields (6.18), which implies that (6.15) holds. Meanwhile, we know from (6.21) that the parameters of an admissible filter are given by (6.19). The proof is completed.

Remark 6.10. Note that in the derivation of [152] and [39], not only a block-diagonal Lyapunov matrix is used, but also it is required that a diagonal block is the same for different Markov modes. This can bring conservatism in the filter design. The main reasons for introducing such block-diagonal Lyapunov matrix rather than the symmetric Lyapunov matrix are:

1. Coupling the parameter matrices of the original systems and the filter with those of estimator of the fault can result from the existence of the non-diagonal block of the latter.
2. The former can make the filter design feasible by introducing less dimension of some auxiliary matrices than one of the latter.

However, the conservatism could be reduced by our proposed approach. That is, the symmetric Lyapunov matrix is introduced and those problems mentioned above are solved by using F_i^T and $\Omega_i F_i$. Thus, our approach could improve the results in [152] and [39].

Remark 6.11. Stability criteria for delayed systems can be classified into two categories: delay-independent and delay-dependent. As time delay is not considered during stability analysis, our results in this paper are delay-independent. That is, our results can be applicable to unknown value of time delay, or even time-varying delay. On the contrary, as time delay is taken into consideration in delay-dependent approach, stability result is less conservative comparatively, especially when the value of time delay is small. To obtain delay-dependent results would be one of our future topics.

Remark 6.12. For discrete-time systems with state delay, one can transform a delayed system into a delay-free system by using state augmentation methods. Stability of delay-free systems can be tested by employing classical results. But delay-free systems will become much complex and thus difficult to analyze with the increase of the size of delays. Moreover, such method is not applicable to time-varying case due to the limitation of available tools. On the contrary, Lyapunov approaches can be particularly good to deal with discrete-time systems with time-varying delay.

6.4 An Illustrative Example

In this section, an example is provided to illustrate the effectiveness of the proposed method.

Consider discrete-time switched system (6.1) consisting of two subsystems with parameters:

$$\begin{aligned}
A_1 &= \begin{bmatrix} 0.2 & -0.1 \\ 0 & 0.4 \end{bmatrix}, A_{d1} = \begin{bmatrix} 0.1 & 0 \\ 0.1 & 0.3 \end{bmatrix}, \\
B_1 &= \begin{bmatrix} 0.2 & 0.1 \end{bmatrix}^T, E_1 = \begin{bmatrix} 0.1 & 0.3 \end{bmatrix}^T, \\
G_1 &= \begin{bmatrix} 1.3 & 1.6 \end{bmatrix}^T, C_1 = \begin{bmatrix} 0.1 & 0 \end{bmatrix}, \\
C_{d1} &= \begin{bmatrix} 0 & 0.1 \end{bmatrix}, \bar{H}_1 = \begin{bmatrix} 0.01 & 0.1 \end{bmatrix}^T, \\
\bar{C}_{11} &= \begin{bmatrix} 0.1 & 0.01 \end{bmatrix}, \bar{C}_{21} = \begin{bmatrix} 0.01 & 0.01 \end{bmatrix}, \\
\bar{C}_{31} &= 0.01, D_1 = 1.1, J_1 = 1.4, Q_1 = 1.0,
\end{aligned}$$

and

$$\begin{aligned}
A_2 &= \begin{bmatrix} 0.4 & 0.1 \\ 0.1 & 0.3 \end{bmatrix}, A_{d2} = \begin{bmatrix} 0.1 & 0 \\ 0.2 & 0.1 \end{bmatrix}, \\
B_2 &= \begin{bmatrix} 0.2 & 0.6 \end{bmatrix}^T, E_2 = \begin{bmatrix} 0.3 & 0.2 \end{bmatrix}^T, \\
G_2 &= \begin{bmatrix} 1.5 & 1.2 \end{bmatrix}^T, C_2 = \begin{bmatrix} 0 & 0.1 \end{bmatrix}, \\
C_{d2} &= \begin{bmatrix} 0.1 & 0 \end{bmatrix}, \bar{H}_1 = \begin{bmatrix} 0.1 & 0.1 \end{bmatrix}^T, \\
\bar{C}_{12} &= \bar{C}_{22} = \begin{bmatrix} 0.1 & 0.1 \end{bmatrix}, \bar{C}_{32} = 0.1, D_2 = 1.2, \\
J_2 &= 1.5, Q_2 = 1.1, \bar{F}(k) = 1, \tau = 1.
\end{aligned}$$

The weighted matrix of the fault is supposed to be $W_f(z) = 0.5z/(z-0.5)$ with the minimal realization:

$$A_w = 0.5, B_w = 0.25, C_w = 1.0, D_w = 0.5.$$

For a given $\gamma = 1.2$, by solving (6.18), we can have a feasible solution. Furthermore, by (6.19), it follows from Theorem 6.9 that:

$$\left[\begin{array}{c|c} A_{f1} & B_{f1} \\ \hline C_{f1} & D_{f1} \end{array}\right] = \left[\begin{array}{cc|c} -0.1065 & -0.4050 & 0.0386 \\ 0.2451 & 0.6430 & -0.2548 \\ \hline -0.2042 & 0.0671 & 0.4552 \end{array}\right],$$

$$\left[\begin{array}{c|c} A_{f2} & B_{f2} \\ \hline C_{f2} & D_{f2} \end{array}\right] = \left[\begin{array}{cc|c} 0.2392 & 0.0403 & -0.0126 \\ -0.0657 & 0.1803 & -0.2144 \\ \hline -0.1760 & -0.0792 & 0.4089 \end{array}\right].$$

An unknown input is assumed to $d_k = 0.01exp(-0.04k)cos(0.03\pi k)$. The control input u_k is the unit step function. The fault signal f_k is simulated as a square wave of unit amplitude occurred from 20 to 40 steps. The switching signal is generated randomly and shown in Fig.6.1. The threshold can be determined as $J_{th} = 4.8983$ for $l_0 = 0$ and $L = 100$. Fig.6.2 shows the evolution of residual evaluation function $J_L(r)$, where the dashed line is fault-free case, the solid line is the case with the fault f_k. The simulation results show that $J_L(r) = 4.9487 > 4.8983$ for $L = 35$, which means that the fault f_k can be detected fifteen time steps after its occurrence.

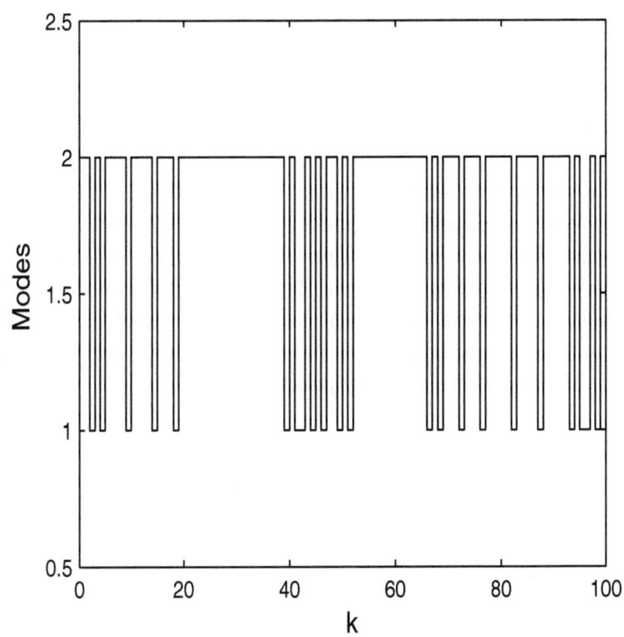

Fig. 6.1 Switching signal

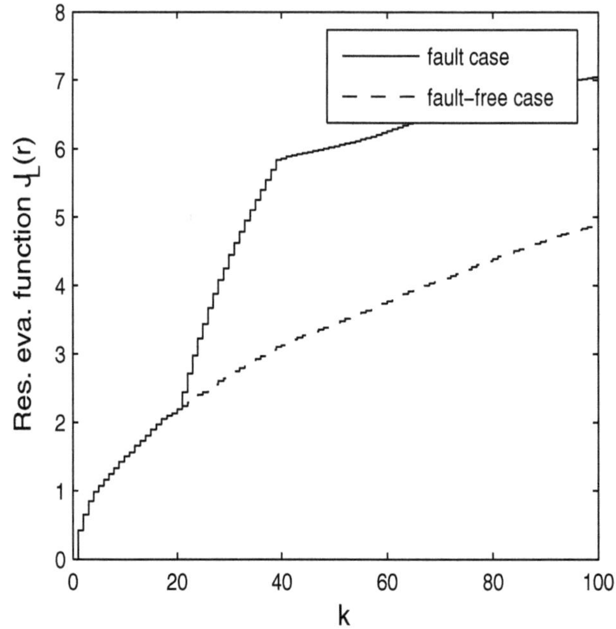

Fig. 6.2 Evolution of residual evaluation function

6.5 Summary

The problem of RFDFD for discrete-time switched systems with state delays has been investigated. A mode-dependent filter is constructed as a residual generator. By augmenting the states of the original system and the filter, the problem of RFDFD has been cast into an H_∞ filtering problem. By using switched Lyapunov functional approach, a sufficient condition for the solvability of this problem is established in terms of LMIs and the desired filter has been constructed. An example has been given to show the effectiveness of the proposed methods.

7

Robust Fault Detection for Continuous-Time Switched Delay Systems

This chapter studies the robust fault detection problem for continuous-time switched systems with state delays. The fault detection filter is used as the residual generator depending on the system mode. Attention is focused on designing the filter such that, for the modeling errors, the unknown inputs and the control inputs, the error between the residuals and the faults is minimized. The addressed fault detection filter design is converted into an auxiliary H_∞ filtering problem. By using the Lyapunov-Krasovskii functional method and average dwell time approach, a sufficient condition for the solvability of this problem is established in terms of linear matrix inequalities (LMIs). Two examples are provided to demonstrate the effectiveness of the proposed method.

7.1 Introduction

As mentioned in the previous chapter, FDI in dynamic systems has been active field of research during the past decades due to an increasing demand for higher safety and reliability standards [21, 73, 84, 22, 47, 86, 23, 14, 39, 151, 12, 80]. Since the unknown inputs and the faults to be detected are coupled and the modeling errors are unavoidable in many industrial systems, FDI has to be sensitive to faults and simultaneously robust to the unknown inputs and the modeling errors, i.e. the robust FDI issues, see for example, [151, 12, 80] and the references therein. Recently, an H_∞-filtering formulation of FDI problem has been presented to solve the robust FDI problem for Markovian jump linear systems [152], linear time-invariant systems [81] and networked control systems [28, 39]. On another research front line, there has been increasing interest in the control problems of switched delay systems due to their significance both in theory and applications (see, for instance, [58, 128, 140] and the references therein). Delay-independent conditions for the existence of a robust fault detection filter for discrete-time switched delay systems under arbitrary switching signals are obtained by using switched Lyapunov functional approach [113]. However, to the best of the authors' knowledge, the problem of FDI for continuous-time switched linear systems with state delays has not been investigated yet. This motivates

D. Wang et al.: *Robust Filtering & Fault Detect. of Switched Delay Systems*, LNCIS 445, pp. 121–135.
DOI: 10.1007/978-3-642-37685-6_7

us to study this interesting and challenging problem, which has great potential in practical applications.

This chapter deals with the problem of robust fault detection filter design by using the Lyapunov-Krasovskii functional method and average dwell time approach for continuous-time switched linear systems with state delays. Firstly, the residual generator is constructed based on the filter and the design of robust fault detection filter is formulated as H_∞ filtering problem. The objective is to make the difference between the faults and the residuals as small as possible, and increase robustness of the residuals to the unknown input and the modeling errors. Secondly, by using the Lyapunov-Krasovskii functional method and average dwell time approach, a sufficient condition on the existence of such filters is established in terms of LMIs. The parameterized matrices of this filter are constructed by solving the corresponding LMIs. Finally, a simulation example is presented to demonstrate the effectiveness of the proposed method.

The rest of this paper is organized as follows. In Section 7.2, system descriptions and definitions are presented. A sufficient condition on the existence of a robust fault detection filter for continuous-time switched systems with state delays is derived in terms of LMIs, and the parameters of the desired filter are constructed by solving the corresponding LMIs in Section 7.3. To demonstrate the validity of the proposed approach, an example is given in Section 7.4 which is followed by a conclusion in Section 7.5.

7.2 Problem Formulation

Consider a class of uncertain switched linear systems with state delay:

$$\begin{aligned} \dot{x}(t) &= A_\sigma x(t) + A_{d\sigma} x(t-d(t)) + E_\sigma u(t) + B_\sigma w(t) + G_\sigma f(t), \\ y(t) &= C_\sigma x(t) + C_{d\sigma} x(t-d(t)) + F_\sigma u(t) + D_\sigma w(t) + J_\sigma f(t), \\ x(\theta) &= \phi(\theta), \quad \theta \in [-h_2, 0], \end{aligned} \tag{7.1}$$

where $x(t) \in R^n$ is the state, $y(t) \in R^r$ is the measured output, $u(t) \in R^p$, $w(t) \in R^q$, $f(t) \in R^s$ are, respectively, the control input, the disturbance input and the fault input which are assumed to belong to $l_2[0,\infty)$. The piecewise constant function σ denoting $\sigma(t)$ for simplicity : $[0, \infty) \to \mathscr{P} = \{1, \cdots, p\}$ is a switching signal to specify, at each time instant t, the index $\sigma \in \mathscr{P}$ of the active subsystem, i.e., $\sigma = i$ means that the ith subsystem is activated. $p > 1$ is the number of subsystems. The ith subsystem is denoted by A_i, A_{di}, E_i, B_i, G_i, C_i, C_{di}, F_i, D_i, J_i, where

$$A_i = \bar{A}_i + \Delta A_i(t),\ A_{di} = \bar{A}_{di} + \Delta A_{di}(t),\ E_i = \bar{E}_i + \Delta E_i(t).$$

$\bar{A}_i$, $\bar{A}_{di}$, $\bar{E}_i$ are known real constant matrices that describe the nominal system. $\Delta A_i(t)$, $\Delta A_{di}(t)$, $\Delta E_i(t)$ are time-varying matrices representing norm-bounded uncertainties with

$$\left[\Delta A_i(t)\ \ \Delta A_{di}(t)\ \ \Delta E_i(t)\right] = H_i \bar{F}(t) \left[\bar{C}_{i1}\ \ \bar{C}_{i2}\ \ \bar{C}_{i3}\right], \tag{7.2}$$

where H_i, $\bar{C}_{i1}$, $\bar{C}_{i2}$, $\bar{C}_{i3}$ are known real constant matrices, while $\bar{F}(t)$ is unknown time-varying matrices satisfying $\bar{F}^T(t)\bar{F}(t) \leq I$. $d(t)$ is assumed to be the interval time-varying delay satisfying either of (A1) and (A2) in the Chapter 5.

Remark 7.1. The matrices H_i, $\bar{C}_{i1}$, $\bar{C}_{i2}$, $\bar{C}_{i3}$ capture the structure of the uncertainty while $\bar{F}(t)$ represents parametric uncertainty. The former specifies how the latter affect the nominal matrices of the system (7.1). The disturbance input $w(t)$ is arbitrary energy-bounded and then it is not necessary to obtain its exact knowledge such as the statistics. The time-varying delay $d(t)$ is differentiable and bounded with a constant delay-derivative bound in (A1) while it is continuous and bounded in (A2). Then, in the sense, the criteria obtained under (A1) are less conservative than those under (A2). However, if the information of the derivation of time delay is unknown, only (A2) can be used to deal with the situation.

An FDI system consists of a residual generator and a residual evaluation stage including an evaluation function and a threshold. For the purpose of residual generation, the following fault detection filter is constructed as a residual generator:

$$\begin{aligned} \dot{\hat{x}}(t) &= A_{f\sigma}\hat{x}(t) + B_{f\sigma}y(t), \\ r(t) &= C_{f\sigma}\hat{x}(t) + D_{f\sigma}y(t), \end{aligned} \tag{7.3}$$

where $\hat{x}(t) \in R^n$ and $r(t) \in R^s$ are the state and the residual, respectively. A_{fi}, B_{fi}, C_{fi} and D_{fi}, $\forall i \in \mathscr{P}$ are the parameterized filter matrices to be determined.

Remark 7.2. The parameters of filter (7.3) depend on the system modes, i.e. the residual generator is a switched system as well. The filter (7.3) is switched synchronously by the switching signal in system (7.1). This means that the switched signal in filter (7.3) is the same as that in system (7.1).

For the purpose of fault detection, it is not necessary to estimate the fault $f(t)$. Sometimes one is more interested in the fault signal of a certain frequency interval, which can be formulated as the weighted fault $\hat{f}(s) = Q_f(s)f(s)$ with $Q_f(s)$ being a given stable weighting matrix. A minimal realization of $\hat{f}(s) = Q_f(s)f(s)$ is supposed to be

$$\begin{aligned} \dot{\bar{x}}(t) &= A_Q\bar{x}(t) + B_Qf(t), \\ \hat{f}(t) &= C_Q\bar{x}(t) + D_Qf(t), \end{aligned} \tag{7.4}$$

where $\bar{x}(t) \in \mathscr{R}^{\bar{n}}$ is the state of the weighted fault, $f(t) \in R^q$ is the original fault and $\hat{f}(t) \in R^q$ is the weighted fault. A_Q, B_Q, C_Q and D_Q are assumed to be known real constant matrices with the appropriate dimensions.

Remark 7.3. Similar to [152] and [39], the introduction of a stable weighting matrix $Q_f(s)$ could limit the frequency ranges of interest, but the system performance can be improved and the frequency characteristics required to reflect the emphases of different frequency ranges can be captured.

Denoting $e(t) = r(t) - \hat{f}(t)$ and augmenting the model of system (7.1) to include the states of (7.3), we can obtain the augmented system as follows:

$$\begin{aligned} \dot{\tilde{x}}(t) &= \tilde{A}_{\sigma}\tilde{x}(t) + \tilde{A}_{d\sigma}K\tilde{x}(t-d(t)) + \tilde{B}_{\sigma}\tilde{w}(t), \\ e(t) &= \tilde{C}_{\sigma}\tilde{x}(t) + \tilde{C}_{d\sigma}K\tilde{x}(t-d(t)) + \tilde{D}_{\sigma}\tilde{w}(t), \end{aligned} \tag{7.5}$$

where

$$\begin{aligned} \tilde{x}(t) &= \begin{bmatrix} x(t) \\ \hat{x}(t) \\ \bar{x}(t) \end{bmatrix}, \ \tilde{w}(t) = \begin{bmatrix} u(t) \\ w(t) \\ f(t) \end{bmatrix}, \\ \tilde{A}_i &= \begin{bmatrix} A_i & 0 & 0 \\ B_{fi}C_i & A_{fi} & 0 \\ 0 & 0 & A_Q \end{bmatrix}, \\ \tilde{A}_{di} &= \begin{bmatrix} A_{di} \\ B_{fi}C_{di} \\ 0 \end{bmatrix}, \ K = \begin{bmatrix} I \\ 0 \\ 0 \end{bmatrix}^T, \\ \tilde{B}_i &= \begin{bmatrix} E_i & B_i & G_i \\ B_{fi}F_i & B_{fi}D_i & B_{fi}J_i \\ 0 & 0 & B_Q \end{bmatrix}, \\ \tilde{C}_i &= \begin{bmatrix} D_{fi}C_i & C_{fi} & -C_Q \end{bmatrix}, \ \tilde{C}_{di} = D_{fi}C_{di}, \\ \tilde{D}_i &= \begin{bmatrix} D_{fi}F_i & D_{fi}D_i & D_{fi}J_i - D_Q \end{bmatrix}. \end{aligned}$$

Now, the Robust Fault Detection Filter (RFDF) design problem to be addressed in this paper can be transformed into an auxiliary H_∞ filtering problem. What we need to do here is to develop the fault detection filter (7.3) for switched delay system (7.1) such that the resulting augmented system (7.5) is exponentially stable under the switching signals with average dwell time and the infimum of γ is made small in the feasibility of

$$\sup_{\|\tilde{w}\|_2 \neq 0} \frac{\|e(t)\|_2}{\|\tilde{w}(t)\|_2} < \gamma, \tag{7.6}$$

for all nonzero $\tilde{w}(t) \in l_2[0,\infty)$ and a given positive scalar γ.

Remark 7.4. It is noted from (7.5) and (7.6) that the residual $r(t)$ generated by the filter (7.3) provides an estimate of the weighted fault $\hat{f}(t)$. The stable weighting matrix $Q_f(s)$ is given. Thus, detection and isolation of the fault $f(t)$ can be achieved by examining the values of the residual $r(t)$. That is, the designed filter not only detects the fault, but also can isolate it. This method to detect the fault has been developed in [39, 81, 152], but it is not extended to uncertain switched delay systems.

After designing the residual generator, the remaining important task is to evaluate the generated residual. One of the widely adopted approaches is to select a threshold and a residual evaluation function. Here, the residual evaluation function is chosen as

$$J_r(L) = \|r(t)\|_{2,L} = \left(\int_0^L r^T(\tau) r(\tau) d\tau \right)^{1/2}, \tag{7.7}$$

where L is the evaluation time window.

Once the evaluation function has been selected, we are able to determine the threshold. It is reasonable to choose the threshold as

$$J_{th} = \sup_{w \in l_2,\, u \in l_2,\, f=0} J_r(T), \tag{7.8}$$

where T is the maximum time window of the evaluation function. It is clear that the computation of J_{th} involves the determination of the unknown inputs $w(t)$ and the control inputs $u(t)$ on the residuals $r(t)$.
Based on this, the faults can be detected by using the following logical relationship:

$$J_r(L) > J_{th} \Rightarrow Faults \Rightarrow Alarm,$$
$$J_r(L) \le J_{th} \Rightarrow No\ Fault.$$

Remark 7.5. In fact, L and T are finite in the horizon since it is desired that the faults will be detected as early as possible, while an evaluation of the residual signals over the whole time range is not practical. This is pointed out in [81, 152, 151, 12].

7.3 Robust Fault Detection Filter Design

In this section, sufficient conditions on the existence of robust fault detection filters can be given. The subsequent lemmas are stated for deriving our results.

Lemma 7.6. *([112]): Given $\alpha > 0$, $\gamma > 0$, $h_2 > h_1 \ge 0$, d and $\mu \ge 1$, the system (7.5) with (A1) under switching signals with average dwell time τ_a is exponentially stable and satisfies (7.6), if there exist symmetric and positive definite matrices P_i, Z_{il}, Q_{ik}, and matrices Y_{ik}, T_{ik}, L_{ik}, $k = 1,2,3, l = 1,2, \forall i \in \mathscr{P}$ such that the following inequalities hold,*

$$P_i \le \mu P_j,\ Q_{ik} \le \mu Q_{jk},\ Z_{il} \le \mu Z_{jl},\ \forall i,\ j \in \mathscr{P}, \tag{7.9}$$

$$\begin{bmatrix} \Omega_i & R_i \\ * & -W_i \end{bmatrix} < 0, \tag{7.10}$$

and average dwell time satisfies

$$\tau_a > \tau_a^* = \frac{\ln \mu}{\alpha}, \tag{7.11}$$

where

$$\Omega_i = \begin{bmatrix} \varphi_{i11} & \varphi_{i12} & \varphi_{i13} \\ * & \varphi_{i22} & \varphi_{i23} \\ * & * & -\gamma^2 I \end{bmatrix},$$

$$R_i = \begin{bmatrix} K^T L_{i1} & -K^T T_{i1} & K^T A_i^T U_i & \tilde{C}_i^T & K^T Y_{i1} & K^T T_{i1} & K^T L_{i1} \\ L_{i2} & -T_{i2} & A_{di}^T U_i & \tilde{C}_{di}^T & Y_{i2} & T_{i2} & L_{i2} \\ L_{i3} & -T_{i3} & \hat{B}_i^T U_i & \tilde{D}_i^T & Y_{i3} & T_{i3} & L_{i3} \end{bmatrix},$$

$$W_i = diag\left\{ e^{-\alpha h_1} Q_{i1},\ e^{-\alpha h_2} Q_{i2},\ U_i,\ I,\ H_{20}^{-1} Z_{i1},\ H_{21}^{-1} Z_{i3},\ H_{21}^{-1} Z_{i2} \right\},$$

$$H_{21} = \frac{e^{\alpha h_2} - e^{\alpha h_1}}{\alpha},\ H_{20} = \frac{e^{\alpha h_2} - 1}{\alpha}, h_{21} = h_2 - h_1,$$

$$U_i = h_2 Z_{i1} + h_{21} Z_{i2},\ Z_{i3} = Z_{i1} + Z_{i2},$$

$$\varphi_{i11} = P_i \tilde{A}_i + \tilde{A}_i^T P_i + \sum_{l=1}^{3} K^T Q_{il} K + K^T Y_{i1} K + K^T Y_{i1}^T K + \alpha P_i,$$

$$\varphi_{i12} = P_i \tilde{A}_{di} - K^T Y_{i1} + K^T Y_{i2}^T + K^T T_{i1} - K^T L_{i1},$$

$$\varphi_{i22} = (d-1) e^{-\alpha h_1} Q_{i3} - Y_{i2} - Y_{i2}^T + T_{i2} + T_{i2}^T - L_{i2} - L_{i2}^T,$$

$$\varphi_{i13} = P_i \tilde{B}_i + K^T Y_{i3}^T,$$

$$\varphi_{i23} = T_{i3}^T - Y_{i3}^T - L_{i3}^T,$$

$$\hat{B}_i = \begin{bmatrix} E_i & B_i & G_i \end{bmatrix}.$$

Lemma 7.7. *([112]): Given $\alpha > 0$, $\gamma > 0$, $h_2 > h_1 \geq 0$, and $\mu \geq 1$, the system (7.5) with (A2) under switching signals with average dwell time is exponentially stable and satisfies (7.6), if there exist symmetric and positive definite matrices P_i, Z_{il}, Q_{il}, and matrices Y_{ik}, T_{ik}, L_{ik}, $k = 1,2,3, l = 1,2, \forall i \in \mathscr{P}$ such that average dwell time τ_a satisfying (7.11) and the following inequalities hold,*

$$P_i \leq \mu P_j,\ Q_{il} \leq \mu Q_{jl},\ Z_{il} \leq \mu Z_{jl},\ \forall i,\ j \in \mathscr{P}, \tag{7.12}$$

$$\begin{bmatrix} \bar{\Omega}_i & R_i \\ * & -W_i \end{bmatrix} < 0, \tag{7.13}$$

where

$$\bar{\Omega}_i = \begin{bmatrix} \bar{\varphi}_{i11} & \varphi_{i12} & \varphi_{i13} \\ * & \bar{\varphi}_{i22} & \varphi_{i23} \\ * & * & -\gamma^2 I \end{bmatrix},$$

$$\bar{\varphi}_{i11} = P_i \tilde{A}_i + \tilde{A}_i^T P_i + \sum_{l=1}^{2} K^T Q_{il} K + K^T Y_{i1} K + K^T Y_{i1}^T K + \alpha P_i,$$

$$\bar{\varphi}_{i22} = -Y_{i2} - Y_{i2}^T + T_{i2} + T_{i2}^T - L_{i2} - L_{i2}^T,$$

and R_i, W_i, φ_{i12}, φ_{i13} and φ_{i23} are defined in Lemma 7.6.

Based on Lemmas 7.6 and 7.7, we will present our main results in this paper as follows.

Theorem 7.8. *For system (7.1) with (A1) and given $\alpha > 0$, $\gamma > 0$, $h_2 > h_1 \geq 0$, d and $\mu \geq 1$, the RFDF problem is solvable, if there exist positive definite and symmetric matrices P_{ik}, Z_{il}, Q_{ik} and matrices Y_{il}, T_{il}, L_{il}, Y_{i3k}, T_{i3k}, L_{i3k}, N_{i1}, N_{i2}, N_{i3}, N_{i4}, $k = 1,2,3, l = 1,2, \forall i \in \mathscr{P}$ such that (7.11) and the following inequalities hold,*

$$P_{i1} - P_{i2} > 0,\ P_{j1} - P_{j2} > 0, \tag{7.14}$$

$$\begin{aligned} &P_{i2} \leq \mu P_{j2},\ P_{i1} - P_{i2} \leq \mu(P_{j1} - P_{j2}), \\ &P_{i3} \leq \mu P_{j3},\ Q_{ik} \leq \mu Q_{jk},\ Z_{il} \leq \mu Z_{jl}, \end{aligned} \tag{7.15}$$

$$\begin{bmatrix} \tilde{\Omega}_i + \Phi_{i1} & \tilde{R}_i & \Phi_{i2} \\ * & -W_i & \Phi_{i3} \\ * & * & -\varepsilon I \end{bmatrix} < 0, \tag{7.16}$$

where

$$\tilde{\Omega}_i = \begin{bmatrix} \tilde{\varphi}_{i11} & \tilde{\varphi}_{i12} & 0 & \tilde{\varphi}_{i14} & \tilde{\varphi}_{i15} & \tilde{\varphi}_{i16} & \tilde{\varphi}_{i17} \\ * & \tilde{\varphi}_{i22} & 0 & \tilde{\varphi}_{i24} & \tilde{\varphi}_{i25} & \tilde{\varphi}_{i26} & \tilde{\varphi}_{i27} \\ * & * & \tilde{\varphi}_{i33} & 0 & 0 & 0 & P_{i3}B_Q \\ * & * & * & \varphi_{i22} & \tilde{\varphi}_{i45} & \tilde{\varphi}_{i46} & \tilde{\varphi}_{i47} \\ * & * & * & * & -\gamma^2 I & 0 & 0 \\ * & * & * & * & * & -\gamma^2 I & 0 \\ * & * & * & * & * & * & -\gamma^2 I \end{bmatrix},$$

$$\tilde{R}_i = \begin{bmatrix} K^T L_{i1} & -K^T T_{i1} & \tilde{\varphi}_{i110} & \hat{C}_i^T & K^T Y_{i1} & K^T T_{i1} & K^T L_{i1} \\ L_{i2} & -T_{i2} & \tilde{\varphi}_{i210} & C_{di}^T N_{i4}^T & Y_{i2} & T_{i2} & L_{i2} \\ L_{i3} & -T_{i3} & \hat{B}_i^T U_i & \hat{D}_i^T & Y_{i3} & T_{i3} & L_{i3} \end{bmatrix},$$

$$\tilde{\varphi}_{i11} = P_{i1}\bar{A}_i + \bar{A}_i^T P_{i1} - N_{i2}C_i - C_i^T N_{i2}^T + \sum_{l=1}^{3} Q_{il} + Y_{i1} + Y_{i1}^T + \alpha P_{i1},$$

$$\tilde{\varphi}_{i12} = C_i^T N_{i2}^T - N_{i1} - \bar{A}_i^T P_{i2} - \alpha P_{i2},$$
$$\tilde{\varphi}_{i22} = N_{i1} + N_{i1}^T + \alpha P_{i2},$$
$$\tilde{\varphi}_{i33} = P_{i3}A_Q + A_Q^T P_{i3} + \alpha P_{i3},$$
$$\tilde{\varphi}_{i14} = P_{i1}\bar{A}_{di} - N_{i2}C_{di} - Y_{i1} + Y_{i2}^T + T_{i1} - L_{i1},$$
$$\tilde{\varphi}_{i24} = N_{i2}C_{di} - P_{i2}\bar{A}_{di},$$
$$\tilde{\varphi}_{i15} = P_{i1}\bar{E}_i - N_{i2}F_i + Y_{i31}^T,$$
$$\tilde{\varphi}_{i25} = N_{i2}F_i - P_{i2}\bar{E}_i,$$
$$\tilde{\varphi}_{i45} = T_{i31}^T - Y_{i31}^T - L_{i31}^T,$$
$$\tilde{\varphi}_{i16} = P_{i1}B_i - N_{i2}D_i + Y_{i32}^T,\ \tilde{\varphi}_{i26} = N_{i2}D_i - P_{i2}B_i,$$
$$\tilde{\varphi}_{i46} = T_{i32}^T - Y_{i32}^T - L_{i32}^T,$$
$$\tilde{\varphi}_{i17} = P_{i1}G_i - N_{i2}J_i + Y_{i33}^T,$$
$$\tilde{\varphi}_{i27} = N_{i2}J_i - P_{i2}G_i,$$
$$\tilde{\varphi}_{i47} = T_{i33}^T - Y_{i33}^T - L_{i33}^T,$$
$$\tilde{\varphi}_{i110}^T = \begin{bmatrix} U_i\bar{A}_i & 0 & 0 \end{bmatrix},\ \tilde{\varphi}_{i210}^T = U_i\bar{A}_{di},$$

$$
\begin{aligned}
Y_{i3}^T &= \begin{bmatrix} Y_{i31}^T \; Y_{i32}^T \; Y_{i33}^T \end{bmatrix}, \\
T_{i3}^T &= \begin{bmatrix} T_{i31}^T \; T_{i32}^T \; T_{i33}^T \end{bmatrix}, \\
L_{i3}^T &= \begin{bmatrix} L_{i31}^T \; L_{i32}^T \; L_{i33}^T \end{bmatrix}, \\
\hat{C}_i &= \begin{bmatrix} N_{i4}C_i \; N_{i3} \; -C_Q \end{bmatrix}, \\
\hat{D}_i &= \begin{bmatrix} N_{i4}F_i \; N_{i4}D_i \; N_{i4}J_i - D_Q \end{bmatrix}, \\
\Phi_{i1} &= \begin{bmatrix} \tilde{C}_{i1}^T\tilde{C}_{i1} & \tilde{C}_{i1}^T\bar{C}_{i2} & \tilde{C}_{i1}^T\tilde{C}_{i3} \\ * & \bar{C}_{i2}^T\bar{C}_{i2} & \bar{C}_{i2}^T\tilde{C}_{i3} \\ * & * & \tilde{C}_{i3}^T\tilde{C}_{i3} \end{bmatrix}, \\
\tilde{C}_{i1} &= \begin{bmatrix} \bar{C}_{i1} \; 0 \; 0 \end{bmatrix}, \\
\tilde{C}_{i3} &= \begin{bmatrix} \bar{C}_{i3} \; 0 \; 0 \end{bmatrix}, \\
\Phi_{i2}^T &= \begin{bmatrix} H_i^T P_{i1} \; -H_i^T P_{i2} \; 0 \; 0 \; 0 \; 0 \; 0 \end{bmatrix}, \\
\Phi_{i3}^T &= \begin{bmatrix} 0 \; 0 \; H_i^T U_i \; 0 \; 0 \; 0 \; 0 \end{bmatrix},
\end{aligned}
$$

and U_i, W_i, φ_{i22} are defined in Lemma 7.6. Moreover, if (7.14), (7.15) and (7.16) have a feasible solution, the RFDF parameter matrices can be constructed by

$$
\begin{aligned}
A_{fi} &= P_{i2}^{-1}N_{i1}, \; B_{fi} = P_{i2}^{-1}N_{i2}, \\
C_{fi} &= N_{i3}, \; D_{fi} = N_{i4},
\end{aligned} \tag{7.17}
$$

and average dwell time is given (7.11).

Proof. By Lemma 7.6, the system (7.5) under switching signals with average dwell time τ_a is exponentially stable and satisfies (7.6), if (7.9), (7.10) and (7.11) hold. We partition $P_i, \forall i \in \mathscr{P}$ as

$$
P_i = \begin{bmatrix} P_{i1} & -P_{i2} & 0 \\ * & P_{i2} & 0 \\ * & * & P_{i3} \end{bmatrix}. \tag{7.18}
$$

Since $P_i > 0$, we have $P_{i1} > 0$, $P_{i2} > 0$, $P_{i3} > 0$ and (7.14).
Note that from (7.9), by Schur complement, we can obtain (7.15).
Defining

$$
\begin{aligned}
N_{i1} &= P_{i2}A_{fi}, \; N_{i2} = P_{i2}B_{fi}, \\
N_{i3} &= C_{fi}, \; N_{i4} = D_{fi}.
\end{aligned} \tag{7.19}
$$

Noticing the uncertainties satisfying (7.2) and substituting (7.19) into (7.10), we have

$$
\begin{bmatrix} \tilde{\Omega}_i & \tilde{R}_i \\ * & -W_i \end{bmatrix} + \tilde{\Xi}_8^T \bar{F}(t)\tilde{\Xi}_9 + \tilde{\Xi}_9^T \bar{F}^T(t)\tilde{\Xi}_8 < 0, \tag{7.20}
$$

where

$$\tilde{\Xi}_8 = \begin{bmatrix} \Phi_{i2}^T & \Phi_{i3}^T \end{bmatrix},$$
$$\tilde{\Xi}_9 = \begin{bmatrix} \tilde{C}_{i1} & \bar{C}_{i2} & \tilde{C}_{i3} & 0 & 0 & 0 & 0 & 0 & 0 & 0 \end{bmatrix}.$$

It follows from Lemma 6.5 that

$$\Phi + \varepsilon \tilde{\Xi}_8^T \tilde{\Xi}_8 + \varepsilon^{-1} \tilde{\Xi}_9^T \tilde{\Xi}_9 < 0. \tag{7.21}$$

Furthermore, by Schur complement, one can get (7.16). Thus, if a solution to LMIs (7.14), (7.15) and (7.16) exists, the parameters of an admissible RFDF are given by (7.17). The proof is completed.

Remark 7.9. We can see from (7.10) that it is difficult to deal with the problem of RFDFD due to the existence of P_i, which leads to some product terms between P_i and A_{fi}, B_{fi}. Therefore, (7.10) is not of strict LMIs. To overcome the difficulties, (7.18) is exploited such that these product terms are decoupled and then sufficient conditions for the solvability of such problems are derived and expressed in terms of strict LMIs. The introduction of a block-diagonal matrix P_i makes it tractable to cope with the problem of RFDFD.

Remark 7.10. If $\mu = 1$, which leads to $P_{il} \equiv P_l$, $Z_{il} \equiv Z_l$, $Q_{ik} \equiv Q_k$, $k = 1,2,3, l = 1,2, \forall i \in \mathscr{P}$ by (7.9) and $\tau^* = 0$ by (7.11), then the augmented system (7.5) possesses a common Lyapunov functional and the switching signals can be arbitrary. This implies that the problem of RFDFD can be solvable under the arbitrary switching signals.

Remark 7.11. In the above criteria, the derivative of the interval time-varying delay is known, i.e. (A1). Unfortunately, in many engineering applications, it is difficult to obtain the information, i.e. (A2). For (A2), Theorem 7.8 is still applicable provided that the terms containing Q_{i3} in Theorem 7.8 are removed. Owe to the limitation of the length of this paper, we only extend Theorem 7.8 to the case (A2) as follows.

Corollary 7.12. *For system (7.1) with (C2) and given $\alpha > 0$, $\gamma > 0$, $h_2 > h_1 \geq 0$ and $\mu \geq 1$, the RFDF problem is solvable, if there exist positive definite and symmetric matrices P_{ik}, Z_{il}, Q_{il} and matrices Y_{il}, T_{il}, L_{il}, Y_{i3k}, T_{i3k}, L_{i3k}, N_{i1}, N_{i2}, N_{i3}, N_{i4}, $k = 1,2,3, l = 1,2, \forall i \in \mathscr{P}$ such that (7.11), (7.14) and the following inequalities hold,*

$$P_{i2} \leq \mu P_{j2},\ P_{i1} - P_{i2} \leq \mu(P_{j1} - P_{j2}),\ P_{i3} \leq \mu P_{j3},$$
$$Q_{il} \leq \mu Q_{jl},\ Z_{il} \leq \mu Z_{jl}, \tag{7.22}$$

$$\begin{bmatrix} \hat{\Omega}_i + \Phi_{i1} & \tilde{R}_i & \Phi_{i2} \\ * & -W_i & \Phi_{i3} \\ * & * & -\varepsilon I \end{bmatrix} < 0, \tag{7.23}$$

where

$$\hat{\Omega}_i = \begin{bmatrix} \hat{\varphi}_{i11} & \tilde{\varphi}_{i12} & 0 & \tilde{\varphi}_{i14} & \tilde{\varphi}_{i15} & \tilde{\varphi}_{i16} & \tilde{\varphi}_{i17} \\ * & \tilde{\varphi}_{i22} & 0 & \tilde{\varphi}_{i24} & \tilde{\varphi}_{i25} & \tilde{\varphi}_{i26} & \tilde{\varphi}_{i27} \\ * & * & \tilde{\varphi}_{i33} & 0 & 0 & 0 & P_{i3}B_Q \\ * & * & * & \bar{\varphi}_{i22} & \tilde{\varphi}_{i45} & \tilde{\varphi}_{i46} & \tilde{\varphi}_{i47} \\ * & * & * & * & -\gamma^2 I & 0 & 0 \\ * & * & * & * & * & -\gamma^2 I & 0 \\ * & * & * & * & * & * & -\gamma^2 I \end{bmatrix},$$

$$\hat{\varphi}_{i11} = P_{i1}\bar{A}_i + \bar{A}_i^T P_{i1} - N_{i2}C_i - C_i^T N_{i2}^T + \sum_{l=1}^{2} Q_{il} + Y_{i1} + Y_{i1}^T + \alpha P_{i1},$$

and $\tilde{R}_i$, Φ_{i1}, Φ_{i2}, Φ_{i3} are defined in Theorem 7.8. Moreover, if (7.14), (7.22) and (7.23) have a feasible solution, the RFDF parameter matrices can be constructed by (7.17).

Proof. By using Lemma 7.7 and the approach of Theorem 7.8, the proof can be easily obtained and thus omitted.

Remark 7.13. There are two methods to obtain μ: one is that we first select a scalar as μ, then solve some LMIs. If these LMIs have no solution, a larger μ will be selected to yield a solution. On the contrary, a smaller μ will be chosen. So, a proper μ could be obtained. The other is that we first solve some LMIs, then find μ according to the obtained solution. This method may not lead to a suitable μ. Therefore, we utilize the former to obtain μ in this paper.

Remark 7.14. Adaptive threshold can be designed as follows: since (7.3) is selected as the residual evaluation function, we get

$$\|r(t)\|_{2,T} = \|r_u(t) + r_w(t) + r_f(t)\|_{2,T}, \tag{7.24}$$

where

$$\begin{aligned} r_u(t) &= r(t)|_{w=0,f=0}, \\ r_w(t) &= r(t)|_{u=0,f=0}, \\ r_f(t) &= r(t)|_{u=0,w=0}. \end{aligned}$$

For the fault-free case, we can select the threshold as

$$\begin{aligned} J_{th} &= J_{th,u} + J_{th,w} \\ &= \sup_{u\in l_2[0,\infty)} \|r_u(t)\|_{2,T} + \sup_{w\in l_2[0,\infty)} \|r_w(t)\|_{2,T} \\ &\geq \|r_u(t) + r_w(t)\|_{2,T}, \end{aligned} \tag{7.25}$$

where $J_{th,w}$ is constant and can be evaluated off-line, while suppose u is known on-line and $J_{th,u}$ can be evaluated real-time by $J_{th,u} = \gamma \|r_u(t)\|_{2,T}$, where

$\gamma = \sup\limits_{u \in l_2[0,\infty)} \frac{\|r_u(t)\|_{2,T}}{\|u(t)\|_{2,T}}$ and can be determined by using Theorem 7.8. This implies that the threshold can be changing by calculated the control input $u(t)$ on-line. In this sense, the threshold is adaptive.

7.4 Illustrative Examples

In this section, two examples are presented to check the validates of the proposed algorithms.

Example 7.15. Consider system (7.1) consisting of two subsystems with parameters: $h_2 = 0.6$, $h_1 = 0.2$, $d = 1.4$,

$$
\begin{aligned}
&\bar{A}_1 = \begin{bmatrix} -3 & 2 \\ 1 & -6 \end{bmatrix}, \bar{A}_{d1} = \begin{bmatrix} -1 & 0 \\ 1 & -1 \end{bmatrix}, \\
&E_1 = \begin{bmatrix} 0 \\ 1 \end{bmatrix}, B_1 = \begin{bmatrix} 0.3 \\ 0.5 \end{bmatrix}, \\
&G_1 = \begin{bmatrix} 0.9 \ 0.7 \end{bmatrix}^T, C_1 = \begin{bmatrix} 1 \ 0 \end{bmatrix}, \\
&C_{d1} = \begin{bmatrix} 0 \ 0 \end{bmatrix}, H_1 = \begin{bmatrix} 0.1 \ 0.01 \end{bmatrix}^T, \\
&\bar{C}_{11} = \begin{bmatrix} 0.01 \ 0.1 \end{bmatrix}, \bar{C}_{12} = \begin{bmatrix} 0.01 \ 0.01 \end{bmatrix}, \\
&J_1 = 0.6, \ Q_1 = 0.8, \\
&D_1 = 0.9, \ \bar{C}_{13} = 0.1,
\end{aligned}
$$

and

$$
\begin{aligned}
&\bar{A}_2 = \begin{bmatrix} -2 & 1 \\ 3 & -9 \end{bmatrix}, \bar{A}_{d2} = \begin{bmatrix} -1 & 0 \\ -1 & -1 \end{bmatrix}, \\
&E_2 = \begin{bmatrix} 1 \\ 0 \end{bmatrix}, B_2 = \begin{bmatrix} 0 \\ 0.1 \end{bmatrix}, \\
&G_2 = \begin{bmatrix} 1 \ 0 \end{bmatrix}^T, C_2 = \begin{bmatrix} 1 \ 0 \end{bmatrix}, \\
&C_{d2} = \begin{bmatrix} 0 \ 0 \end{bmatrix}, H_2 = \begin{bmatrix} 0.1 \ 0.1 \end{bmatrix}^T, \\
&\bar{C}_{21} = \begin{bmatrix} 0.1 \ 0.01 \end{bmatrix}, \bar{C}_{22} = \begin{bmatrix} 0.01 \ 0.01 \end{bmatrix}, \\
&D_2 = 0.9, \ J_2 = 0.4, \\
&Q_2 = 0.9, \ \bar{C}_{23} = 0.1.
\end{aligned}
$$

The weighted matrix of the fault is supposed to be $Q_f(s) = 5/(s+5)$ with the minimal realization:

$$A_Q = -5, \ B_Q = 5, \ C_Q = 1.0, \ D_Q = 0.$$

To demonstrate the effectiveness of the design, an unknown input $d(t)$ is assumed to $d(t) = 0.1exp(-0.04t)cos(0.3\pi t)$. The control input $u(t)$ is the unit step function. The fault signal $f(t)$ is simulated as a square wave of unit amplitude occurred from

6 to 8(s). Hence, given $\alpha = 0.5$, $\mu = 1.2$ and $\gamma = 1.1$, by utilizing the LMI Toolbox, it follows from Theorem 7.8 that we can have a feasible solution. Furthermore, by (7.17), the desired filter can be obtained with the parameterized matrices as follows:

$$\left[\begin{array}{c|c} A_{1f} & B_{1f} \\ \hline C_{1f} & D_{1f} \end{array}\right] = \left[\begin{array}{cc|c} -7.1413 & 3.2672 & 3.0186 \\ 0.1400 & -13.2288 & 2.9322 \\ \hline -0.4156 & 0.1619 & 0.3638 \end{array}\right],$$

and

$$\left[\begin{array}{c|c} A_{2f} & B_{2f} \\ \hline C_{2f} & D_{2f} \end{array}\right] = \left[\begin{array}{cc|c} -7.0836 & 1.1351 & 4.1463 \\ -2.6643 & -15.2569 & 6.7083 \\ \hline -0.3462 & 0.0092 & 0.3433 \end{array}\right],$$

and the minimal average dwell time $\tau_a^* = 0.3646$ is given by (7.11). This means that a desired filter under the switching signals with average dwell time is developed by the proposed method. The switching signal is generated by choosing average dwell time $\tau_a = 0.5$ shown in Fig. 7.1. The threshold can be determined as $J_{th} = 2.5942$ for $T = 20s$. Fig. 7.2 shows the evolution of residual evaluation function $J_r(t)$, in which the dashed line is fault-free case, the solid line is the case with the fault $f(t)$. The simulation results show that $J_r(t) = 2.6238 > 2.5942$ when $t = 12s$, which means that the fault $f(t)$ can be detected six seconds after its occurrence.

Example 7.16. As for (A2), consider system (7.1) with the parameters which are the same as those in Example 7.15. By Corollary 7.12, we can obtain the parameters of the robust fault detection filter as follow:

$$\left[\begin{array}{c|c} A_{1f} & B_{1f} \\ \hline C_{1f} & D_{1f} \end{array}\right] = \left[\begin{array}{cc|c} -7.2129 & 3.1422 & 3.0602 \\ 0.3973 & -13.3617 & 2.8799 \\ \hline -0.4206 & 0.1674 & 0.3624 \end{array}\right],$$

and

$$\left[\begin{array}{c|c} A_{2f} & B_{2f} \\ \hline C_{2f} & D_{2f} \end{array}\right] = \left[\begin{array}{cc|c} -7.0551 & 0.9910 & 4.1858 \\ -2.6116 & -15.1431 & 6.6492 \\ \hline -0.3469 & 0.0106 & 0.3435 \end{array}\right],$$

and the minimal average dwell time $\tau_a^* = 0.3646$ is given by (7.11). Average dwell time is still selected as $\tau_a = 0.5$. The switching signal is shown in Fig. 7.3. The

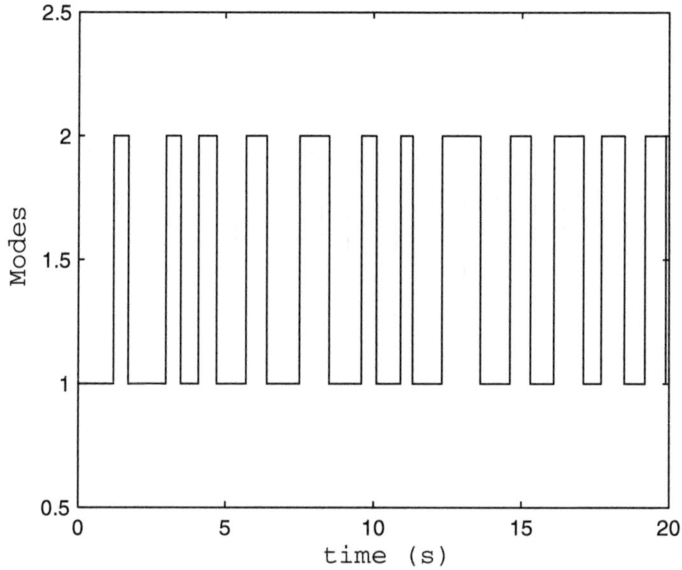

Fig. 7.1 Residual signals for (A1)

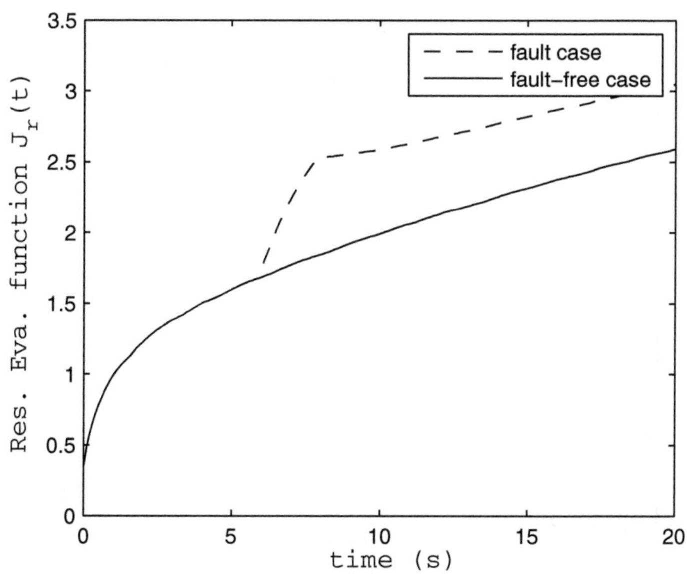

Fig. 7.2 Evolution of residual evaluation function for (A1)

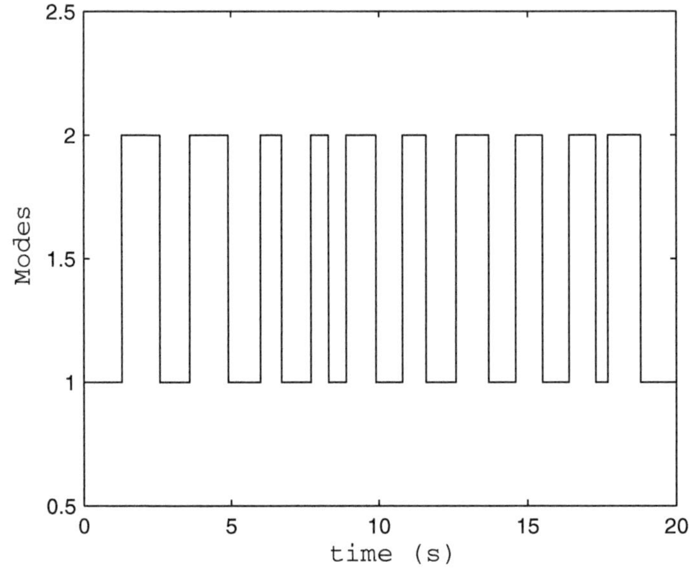

Fig. 7.3 Residual signals for (A2)

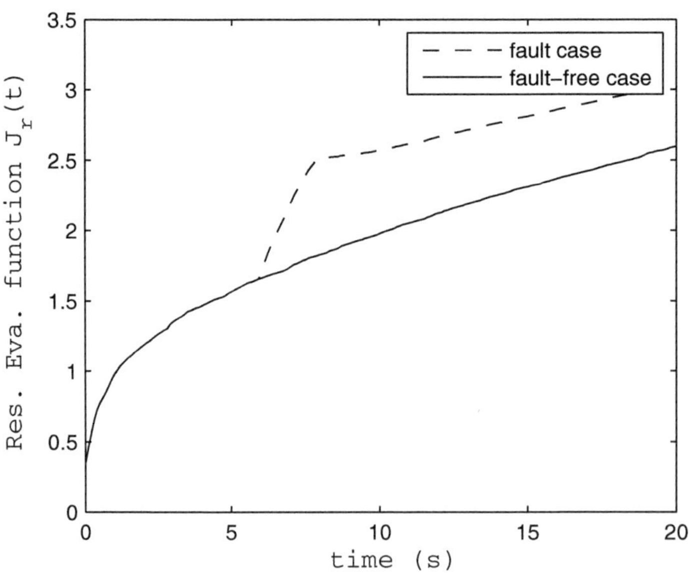

Fig. 7.4 Evolution of residual evaluation function for (A2)

threshold can be determined as $J_{th} = 2.5664$ for $T = 20s$. Fig. 7.4 shows the evolution of residual evaluation function $J_r(t)$, in which the dashed line is fault-free case, the solid line is the case with the fault $f(t)$. The simulation results show that $J_r(t) = 2.5869 > 2.5664$ when $t = 11s$, which means that the fault $f(t)$ can be detected five seconds after its occurrence.

7.5 Summary

In this paper, we have investigated the problem of robust fault detection filter design for continuous-time uncertain switched systems with state delays. The filters are dependent on the original switched systems, i.e. the switching signals are the same. The suitable filters have been constructed under the switching signals with average dwell time, which guarantee sensitivity to the occurred fault and robustness to the modeling errors, the control inputs and the unknown inputs. Two examples have been provided to illustrate the effectiveness of the proposed methods.

8

Conclusions and Perspectives

8.1 Conclusions

Many practical systems can be modeled as switched delay systems and the existence of switching and a time delay brings new challenges. The introduction of switching mechanism leads to complicated system behaviors. Even though the subsystems are linear delay systems, switched delay systems could exhibit nonlinear characteristics under different switching signals. Hence, it is of great theoretical and practical application significance to analyze and synthesize of switched delay systems. This book focuses on the issues of H_∞ filter design and robust fault detection of switched delay systems. The main contributions presented in this book include

(1) Based on switched Lyapunov function method, delay-independent sufficient conditions on the existence of H_∞ filter for switched linear systems with a constant delay are derived and a linearization technique is developed to construct such filter. Furthermore, by using free-weighting matrix approach, delay-independent sufficient conditions are improved to delay-dependent sufficient conditions. Meanwhile, a time delay under consideration is interval time-varying delay.

(2) A novel Lyapunov-Krasovskii function with a weighted factor is proposed and using ADT approach, sufficient conditions for the solvability of exponential H_∞ filtering problem of discrete-time switched delay systems are derived, which are dependent on the upper and lower bounds of the time-varying delay. Based on the exponential H_∞ criterion, a new linearization technique is presented to establish sufficient conditions on the existence of reduced-order model for discrete-time switched delay systems in term of strict LMI to lessen the computation complexity.

(3) By utilizing average dwell time method and linear decoupling techniques, sufficient conditions on the existence of exponential H_∞ filter are presented. The time delay under consideration contains two cases: one is that the time delay is differentiable and bounded with a constant delay-derivative bound while the

D. Wang et al.: *Robust Filtering & Fault Detect. of Switched Delay Systems*, LNCIS 445, pp. 137–139.
DOI: 10.1007/978-3-642-37685-6_8

other is that the time delay is continuous and bounded. Furthermore, the criterion is extended to the problem of model reductions for such switched systems.

(4) The fault detection filter is selected as the residual generator. Its objective is to minimize the estimation error between the residuals and the faults for unknown inputs, control inputs and model uncertainties. By switched Lyapunov functional approach, a sufficient condition for the solvability of the fault detection problem for discrete-time switched delay systems is established, which is delay-independent. Moreover, robust fault detection problem for continuous-time switched systems with state delays is converted into an auxiliary H_∞ filtering problem. By using the weighting Lyapunov-Krasovskii functional method and average dwell time approach, a sufficient condition for the solvability of this problem is established.

8.2 Perspectives

Generally speaking, the theory of switched delay systems is far from well established and in the early stage of development. There still exist a great number of crucial and fundamental issues to be exploited, which are open and challenging. To the best of our knowledge, some problems are listed as follows.

(1) The design of the switching law plays vital roles in stability analysis and synthesis of switched delay systems.

This book studies arbitrary switching laws or average dwell time switching laws. These switching laws are dependent on the time. Sometimes, it requires to design state-dependent switching signals to meet the system performance index such as stabilization of switched delay systems with unstable subsystems [64, 150]. In fact, such stabilization leads to the chattering near the origin. An effective method is to combine with the dwell time technique. In these design switching laws, the system filter design, fault detection and fault tolerance are of more practical significance for further investigation.

(2) A time delay emerges in the control input and the switching signal.

This book studies a class of switched systems with time delays in the subsystem state. However, it is possible to involve time delays in the control input of the system [109, 68]. For instance, in the remote control, the control input signal is transmitted through the network where a time delay is inevitable. In the control of the real systems, it is difficult to accurately obtain and transmit a switching signal and thus the switching signals may contain a time delay. Asynchronous switching is a representative example in which a time delay is in a switching signal.

(3) There is still a bottleneck to analyze and synthesize nonlinear switched delay systems.

This book studies the linear subsystem model, which has displayed the essentially nonlinear features. However, in practice, many subsystems could be nonlinear. Although the nonlinear subsystems can be locally linearized, the results obtained are far from ideal. So far, only a method, common Lyapunov function approach, is applicable. As is well known, this approach is very conservative. Therefore, how to make use of the theory of nonlinear systems and tools in traditional control and system field to analyze nonlinear switched delay systems is a great challenging [13, 63].

(4) Switched delay systems possess random information in the subsystems.

Switched stochastic systems are of practical significance due to the random behavior of internal components and external environment in the real system. For example, Markov jump linear delay systems can model a class of stochastic characteristics, but they require that the occurrences of switching time instant obey the Markov process [149]. In the dynamic systems, the stochastic noise is introduced to comprise another kind of stochastic switched delay systems [124, 20]. However, how to model a more general stochastic switched delay consisting in the random in time delays, switching signals, dynamics and so on, is of further exploration.

(5) The application of the switched delay systems is difficult.

In theory, switched delay systems have a strong engineering background, especially in the network control systems, chemical engineering, and electrical systems [110, 103]. The common features of these system models include inevitable practical problems such as switch on, switch off, signal transmission delay, etc. But in real engineering, there are a great deal of issues in need of being taken into account. If switched delay systems theory is applied to the real system, it will have a tremendous economic benefits.

References

1. Anderson, B., Moore, J.: Optimal filtering, vol. 11. Prentice-Hall, Englewood Cliffs (1979)
2. Antsaklis, P.: A brief introduction to the theory and applications of hybrid systems. Proceedings of the IEEE 88(6), 887–897 (2000)
3. Basin, M., Rodkina, A.: On delay-dependent stability for a class of nonlinear stochastic systems with multiple state delays. Nonlinear Analysis: Theory, Methods and Applications 68(8), 2147–2157 (2008)
4. Beard, R.: Failure accommodation in linear systems through self-reorganization. Ph.D. thesis, Massachusetts Institute of Technology (1971)
5. Boyd, S., El Ghaoul, L., Feron, E., Balakrishnan, V.: Linear matrix inequalities in system and control theory, vol. 15. Society for Industrial Mathematics (1987)
6. Branicky, M.: Multiple lyapunov functions and other analysis tools for switched and hybrid systems. IEEE Transactions on Automatic Control 43(4), 475–482 (1998)
7. Casavola, A., Famularo, D., Franzè, G.: A robust deconvolution scheme for fault detection and isolation of uncertain linear systems: an lmi approach. Automatica 41(8), 1463–1472 (2005)
8. Chen, J., Patton, R.: Robust model-based fault diagnosis for dynamic systems. Kluwer Academic Publishers (1999)
9. Chiang, L., Russell, E., Braatz, R.: Fault detection and diagnosis in industrial systems. Springer (2001)
10. Daafouz, J., Riedinger, P., Iung, C.: Stability analysis and control synthesis for switched systems: a switched lyapunov function approach. IEEE Transactions on Automatic Control 47(11), 1883–1887 (2002)
11. DeCarlo, R., Branicky, M., Pettersson, S., Lennartson, B.: Perspectives and results on the stability and stabilizability of hybrid systems. Proceedings of the IEEE 88(7), 1069–1082
12. Ding, S., Jeinsch, T., Frank, P., Ding, E.: A unified approach to the optimization of fault detection systems. International Journal of Adaptive Control and Signal Processing 14(7), 725–745 (2000)
13. Dong, Y., Liu, J., Mei, S., Li, M.: Stabilization for switched nonlinear time-delay systems. Nonlinear Analysis: Hybrid Systems 5(1), 78–88 (2011)
14. Douglas, R., Speyer, J.: h_∞ bounded fault detection filter. Journal of Guidance, Control, and Dynamics 22(1), 129–138 (1999)

15. Du, D., Jiang, B., Shi, P., Zhou, S.: h_∞ filtering of discrete-time switched systems with state delays via switched lyapunov function approach. IEEE Transactions on Automatic Control 52(8), 1520–1525 (2007)
16. Duan, Z., Zhang, J., Zhang, C., Mosca, E.: Robust h_2 and h_∞ filtering for uncertain linear systems. Automatica 42(11), 1919–1926 (2006)
17. Dugard, L., Verriest, E.: Stability and control of time-delay systems. Springer, London (1998)
18. El Ghaoui, L., Oustry, F., AitRami, M.: A cone complementarity linearization algorithm for static output-feedback and related problems. IEEE Transactions on Automatic Control 42(8), 1171–1176 (1997)
19. Elsayed, A., Grimble, M.: A new approach to the h_∞ design of optimal digital linear filters. IMA Journal of Mathematical Control and Information 6(2), 233–251 (1989)
20. Feng, W., Tian, J., Zhao, P.: Stability analysis of switched stochastic systems. Automatica 47(1), 148–157 (2011)
21. Frank, P.: Fault diagnosis in dynamic systems using analytical and knowledge-based redundancy: A survey and some new results. Automatica 26(3), 459–474 (1990)
22. Frank, P.: Enhancement of robustness in observer-based fault detection. International Journal of Control 59(4), 955–981 (1994)
23. Frank, P., Ding, X.: Survey of robust residual generation and evaluation methods in observer-based fault detection systems. Journal of Process Control 7(6), 403–424 (1997)
24. Fridman, E., Shaked, U.: A descriptor system approach to h_∞ control of linear time-delay systems. IEEE Transactions on Automatic Control 47(2), 253–270 (2002)
25. Fridman, E., Shaked, U.: Delay-dependent stability and h_∞ control: constant and time-varying delays. International Journal of Control 76(1), 48–60 (2003)
26. Fu, M., Luo, Z.: Computational complexity of a problem arising in fixed order output feedback design. Systems and Control letters 30(5), 209–215 (1997)
27. Gao, H., Chen, T.: New results on stability of discrete-time systems with time-varying state delay. IEEE Transactions on Automatic Control 52(2), 328–334 (2007)
28. Gao, H., Chen, T., Wang, L.: Robust fault detection with missing measurements. International Journal of Control 81(5), 804–819 (2008)
29. Gao, H., Lam, J., Wang, C., Xu, S.: h_∞ model reduction for discrete time-delay systems: delay-independent and dependent approaches. International Journal of Control 77(4), 321–335 (2004)
30. Gao, H., Wang, C.: A delay-dependent approach to robust H_∞ filtering for uncertain discrete-time state-delayed systems. IEEE Transactions on Signal Processing 52(6), 1631–1640 (2004)
31. Glover, K.: All optimal hankel-norm approximations of linear multivariable systems and their l_∞-error bounds. International Journal of Control 39(6), 1115–1193 (1984)
32. Glover, K., McFarlane, D.: Robust stabilization of normalized coprime factor plant descriptions with h_∞-bounded uncertainty. IEEE Transactions on Automatic Control 34(8), 821–830 (1989)
33. Gopalsamy, K., He, X.: Delay-independent stability in bidirectional associative memory networks. IEEE Transactions on Neural Networks 5(6), 998–1002 (1994)
34. Grigoriadis, K., Watson Jr., J.: Reduced-order H_∞ and $l_2 - l_\infty$ filtering via linear matrix inequalities. IEEE Transactions on Aerospace and Electronic Systems 33(4), 1326–1338 (1997)
35. Grimble, M.: Polynomial systems approach to optimal linear filtering and prediction. International Journal of Control 41(6), 1545–1564 (1985)

36. Grimble, M., El Sayed, A.: Solution of the h_∞ optimal linear filtering problem for discrete-time systems. IEEE Transactions on Acoustics, Speech and Signal Processing 38(7), 1092–1104 (1990)
37. Gu, K., Kharitonov, V., Chen, J.: Stability of time-delay systems. Birkhauser (2003)
38. Gugercin, S., Antoulas, A., Beattie, C.: h_2 model reduction for large-scale linear dynamical systems. SIAM Journal on Matrix Analysis and Applications 30(2), 609–638 (2008)
39. He, X., Wang, Z., Ji, Y., Zhou, D.: Network-based fault detection for discrete-time state-delay systems: A new measurement model. International Journal of Adaptive Control and Signal Processing 22(5), 510–528 (2008)
40. He, Y., Wang, Q., Lin, C., Wu, M.: Delay-range-dependent stability for systems with time-varying delay. Automatica 43(2), 371–376 (2007)
41. He, Y., Wu, M., She, J., Liu, G.: Parameter-dependent Lyapunov functional for stability of time-delay systems with polytopic-type uncertainties. IEEE Transactions on Automatic Control 49(5), 828–832 (2004)
42. Hespanha, J., Morse, A.: Stability of switched systems with average dwell-time. In: Proceedings of the 38th IEEE Conference on Decision and Control, vol. 3, pp. 2655–2660 (1999)
43. Hespanha, J., Morse, A.: Switching between stabilizing controllers. Automatica 38(11), 1905–1917 (2002)
44. Hetel, L., Daafouz, J., Iung, C.: Stabilization of arbitrary switched linear systems with unknown time-varying delays. IEEE Transactions on Automatic Control 51(10), 1668–1674 (2006)
45. Hou, L., Michel, A., Ye, H.: Stability analysis of switched systems. In: Proceedings of the 35th IEEE Decision and Control, vol. 2, pp. 1208–1212 (1996)
46. Hu, K., Yuan, J.: Improved robust h_∞ filtering for uncertain discrete-time switched systems. IET Control Theory and Applications 3(3), 315–324 (2009)
47. Isermann, R.: Process fault detection based on modeling and estimation methods–a survey. Automatica 20(4), 387–404 (1984)
48. Isermann, R.: Fault-diagnosis systems: an introduction from fault detection to fault tolerance. Springer (2005)
49. Isermann, R., Balle, P.: Trends in the application of model-based fault detection and diagnosis of technical processes. Control Engineering Practice 5(5), 709–719 (1997)
50. Ishii, H., Francis, B.: Stabilizing a linear system by switching control with dwell time. IEEE Transactions on Automatic Control 47(12), 1962–1973 (2002)
51. Iwasaki, T., Skelton, R.: All controllers for the general h_∞ control problem: Lmi existence conditions and state space formulas. Automatica 30(8), 1307–1317 (1994)
52. Jaulin, L.: Applied interval analysis: with examples in parameter and state estimation, robust control and robotics. Springer (2001)
53. Jiang, X., Han, Q., Yu, X.: Stability criteria for linear discrete-time systems with interval-like time-varying delay. In: Proceedings of the 2005 American Control Conference, pp. 2817–2822 (2005)
54. Kalman, R.: A new approach to linear filtering and prediction problems. Journal of Basic Engineering 82(1), 35–45 (1960)
55. Kalman, R., Bucy, R.: New results in linear filtering and prediction theory. Journal of Basic Engineering 83(3), 95–108 (1961)
56. Kharitonov, V.: Robust stability analysis of time delay systems: A survey. Annual Reviews in Control 23, 185–196 (1999)
57. Kharitonov, V., Zhabko, A.: Lyapunov–krasovskii approach to the robust stability analysis of time-delay systems. Automatica 39(1), 15–20 (2003)

58. Kim, S., Campbell, S., Liu, X.: Stability of a class of linear switching systems with time delay. IEEE Transactions on Circuits and Systems I: Regular Papers 53(2), 384–393 (2006)
59. Kyoo Kim, D., Park, P., Ko, J.: Output-feedback h_∞ control of systems over communication networks using a deterministic switching system approach. Automatica 40(7), 1205–1212 (2004)
60. Lastman, G.: Reduced-order aggregated models for bilinear time-invariant dynamical systems. IEEE Transactions on Automatic Control 29(4), 359–361 (1984)
61. Leibfritz, F.: An lmi-based algorithm for designing suboptimal static $h_2 \backslash h_\infty$ output feedback controllers. SIAM Journal on Control and Optimization 39(6), 1711–1735 (2001)
62. Li, H., Fu, M.: A linear matrix inequality approach to robust h_∞ filtering. IEEE Transactions on Signal Processing 45(9), 2338–2350 (1997)
63. Lian, J., Wang, M.: Sliding-mode control of switched delay systems with nonlinear perturbations: average dwell time approach. Nonlinear Dynamics 62(4), 791–798 (2010)
64. Lian, J., Zhao, J.: Adaptive variable structure control for uncertain switched delay systems. Circuits, Systems, and Signal Processing 29(6), 1089–1102 (2010)
65. Liberzon, D.: Switching in systems and control. Springer (2003)
66. Liberzon, D., Hespanha, J., Morse, A.: Stability of switched systems: a lie-algebraic condition. Systems and Control Letters 37(3), 117–122 (1999)
67. Liberzon, D., Morse, A.: Basic problems in stability and design of switched systems. IEEE Control Systems Magazine 19(5), 59–70 (1999)
68. Lien, C.H., Yu, K.W., Chung, Y.J., Chang, H.C., Chen, J.D.: Switching signal design for global exponential stability of uncertain switched nonlinear systems with time-varying delay. Nonlinear Analysis: Hybrid Systems 5(1), 10–19 (2011)
69. Lin, H., Antsaklis, P.: Stability and persistent disturbance attenuation properties for a class of networked control systems: switched system approach. International Journal of Control 78(18), 1447–1458 (2005)
70. Lin, H., Antsaklis, P.: Stability and stabilizability of switched linear systems: a survey of recent results. IEEE Transactions on Automatic control 54(2), 308–322 (2009)
71. Liu, X.: Stability analysis of switched positive systems: A switched linear copositive lyapunov function method. IEEE Transactions on Circuits and Systems II: Express Briefs 56(5), 414–418 (2009)
72. Mahmoud, M.: Switched delay-dependent control policy for water-quality systems. IET Control Theory and Applications 3(12), 1599–1610 (2009)
73. Mehra, R., Peschon, J.: An innovations approach to fault detection and diagnosis in dynamic systems. Automatica 7(5), 637–640 (1971)
74. Meng, X., Lam, J., Du, B., Gao, H.: A delay-partitioning approach to the stability analysis of discrete-time systems. Automatica 46(3), 610–614 (2010)
75. Moir, T., Grimble, M.: Optimal self-tuning filtering, prediction, and smoothing for discrete multivariable processes. IEEE Transactions on Automatic Control 29(2), 128–137 (1984)
76. Moon, Y., Park, P., Kwon, W., Lee, Y.: Delay-dependent robust stabilization of uncertain state-delayed systems. International Journal of Control 74(14), 1447–1455 (2001)
77. Moore, B.: Principal component analysis in linear systems: Controllability, observability, and model reduction. IEEE Transactions on Automatic Control 26(1), 17–32 (1981)
78. Morse, A.: Supervisory control of families of linear set-point controllers part i. exact matching. IEEE Transactions on Automatic Control 41(10), 1413–1431 (1996)
79. Narendra, K., Balakrishnan, J.: A common lyapunov function for stable lti systems with commuting a-matrices. IEEE Transactions on Automatic Control 39(12), 2469–2471 (1994)

80. Nguang, S., Shi, P., Ding, S.: Fault detection for uncertain fuzzy systems: An lmi approach. IEEE Transactions on Fuzzy Systems 15(6), 1251–1262 (2007)
81. Nobrega, E., Abdalla, M., Grigoriadis, K.: Lmi-based filter design for fault detection and isolation. In: Proceedings of the 39th IEEE Conference on Decision and Control, vol. 5, pp. 4329–4334 (2000)
82. de Oliveira, M., Bernussou, J., Geromel, J.: A new discrete-time robust stability condition. Systems and Control Letters 37(4), 261–265 (1999)
83. Ooba, T., Funahashi, Y.: On a common quadratic lyapunov function for widely distant systems. IEEE Transactions on Automatic Control 42(12), 1697–1699 (1997)
84. Park, J., Halevi, Y., Rizzoni, G.: A new interpretation of the fault-detection filter part 2: The optimal detection filter. International Journal of Control 60(6), 1339–1351 (1994)
85. Park, P.: A delay-dependent stability criterion for systems with uncertain time-invariant delays. IEEE Transactions on Automatic Control 44(4), 876–877 (1999)
86. Patton, R., Chen, J.: Observer-based fault detection and isolation: robustness and applications. Control Engineering Practice 5(5), 671–682 (1997)
87. Patton, R., Frank, P.: Issues of fault diagnosis for dynamic systems. Springer (2000)
88. Peleties, P., DeCarlo, R.: Asymptotic stability of m-switched systems using lyapunov-like functions. In: 1991 American Control Conference, pp. 1679–1684 (1991)
89. Qiu, J., Feng, G., Yang, J.: Improved delay-dependent h_∞ filtering design for discrete-time polytopic linear delay systems. IEEE Transactions on Circuits and Systems II: Express Briefs 55(2) (2008)
90. Qiu, J., Feng, G., Yang, J.: New results on robust energy-to-peak filtering for discrete-time switched polytopic linear systems with time-varying delay. IET Control Theory and Applications 2(9), 795–806 (2008)
91. Qiu, J., Feng, G., Yang, J.: Robust mixed h_2/h_∞ filtering design for discrete-time switched polytopic linear systems. IET Control Theory and Applications 2(5), 420–430 (2008)
92. Reif, K., Unbehauen, R.: The extended kalman filter as an exponential observer for nonlinear systems. IEEE Transactions on Signal Processing 47(8), 2324–2328 (1999)
93. Richard, J.: Time-delay systems: an overview of some recent advances and open problems. Automatica 39(10), 1667–1694 (2003)
94. Rodkina, A., Basin, M.: On delay-dependent stability for vector nonlinear stochastic delay-difference equations with volterra diffusion term. Systems & Control Letters 56(6), 423–430 (2007)
95. Shaked, U.: h_∞-minimum error state estimation of linear stationary processes. IEEE Transactions on Automatic Control 35(5), 554–558 (1990)
96. Shi, P., Boukas, E., Agarwal, R.: Control of markovian jump discrete-time systems with norm bounded uncertainty and unknown delay. IEEE Transactions on Automatic Control 44(11), 2139–2144 (1999)
97. Shorten, R., Narendra, K.: On the stability and existence of common lyapunov functions for stable linear switching systems. In: Proceedings of the 37th IEEE Conference on Decision and Control, vol. 4, pp. 3723–3724. IEEE (1998)
98. Shorten, R., Narendra, K.: Necessary and sufficient conditions for the existence of a common quadratic lyapunov function for a finite number of stable second order linear time-invariant systems. International Journal of Adaptive Control and Signal Processing 16(10), 709–728 (2002)
99. Shuzhi, S., Sun, Z.: Switched linear systems: Control and design. Springer (2005)
100. Simon, D.: Optimal state estimation: Kalman, H_∞ and nonlinear approaches. Wiley-Interscience (2006)

101. Skafidas, E., Evans, R.J., Savkin, A.V., Petersen, I.R.: Stability results for switched controller systems. Automatica 35(4), 553–564 (1999)
102. Spanos, J., Milman, M., Mingori, D.: A new algorithm for l_2 optimal model reduction. Automatica 28(5), 897–909 (1992)
103. Sun, G., Wang, M., Yao, X., Wu, L.: Fault detection of switched linear systems with its application to turntable systems. Journal of Systems Engineering and Electronics 22(1), 120–126 (2011)
104. Sun, X., Liu, G., Rees, D., Wang, W.: Stability of systems with controller failure and time-varying delay. IEEE Transactions on Automatic Control 53(10), 2391–2396 (2008)
105. Sun, X., Zhao, J., Hill, D.: Stability and l_2-gain analysis for switched delay systems: A delay-dependent method. Automatica 42(10), 1769–1774 (2006)
106. Sun, Z.: Switched linear systems: Control and design. Springer (2005)
107. Sun, Z., Ge, S.: Analysis and synthesis of switched linear control systems. Automatica 41(2), 181–195 (2005)
108. Tjärnström, F., Ljung, L.: l_2 model reduction and variance reduction. Automatica 38(9), 1517–1530 (2002)
109. Vu, L., Morgansen, K.A.: Stability of time-delay feedback switched linear systems. IEEE Transactions on Automatic Control 55(10), 2385–2390 (2010)
110. Wang, D., Shi, P., Wang, W., Karimi, H.R.: Non-fragile h_∞ control for switched stochastic delay systems with application to water quality process. International Journal of Robust and Nonlinear Control (2013, early view)
111. Wang, D., Wang, W., Shi, P.: Delay-dependent exponential stability for switched delay systems. Optimal Control Applications and Methods 30(4), 383–397 (2009)
112. Wang, D., Wang, W., Shi, P.: Exponential h_∞ filtering for switched linear systems with interval time-varying delay. International Journal of Robust and Nonlinear Control 19(5), 532–551 (2009)
113. Wang, D., Wang, W., Shi, P.: Robust fault detection for switched linear systems with state delays. IEEE Transactions on Systems, Man, and Cybernetics, Part B: Cybernetics 39(3), 800–805 (2009)
114. Wang, D., Wang, W., Shi, P., Sun, X.: Controller failure analysis for systems with interval time-varying delay: a switched method. Circuits, Systems, and Signal Processing 28(3), 389–407 (2009)
115. Wang, Y., Yang, G.: h_∞ control of networked control systems with time delay and packet disordering. IET Control Theory and Applications 1(5), 1344–1354 (2007)
116. Weiner, N.: Extrapolation, interpolation, and smoothing of stationary time series with engineering applications. MIT press (1949)
117. Wicks, M., Peleties, P., DeCarlo, R.: Switched controller synthesis for the quadratic stabilisation of a pair of unstable linear systems. European Journal of Control 4(2), 140–147 (1998)
118. Wicks, M.A., Peleties, P., DeCarlo, R.A.: Construction of piecewise lyapunov functions for stabilizing switched systems. In: Proceedings of the 33rd IEEE Conference on Decision and Control, vol. 4, pp. 3492–3497 (1994)
119. Willsky, A.: A survey of design methods for failure detection in dynamic systems. Automatica 12(6), 601–611 (1976)
120. Wu, L., Lam, J.: Sliding mode control of switched hybrid systems with time-varying delay. International Journal of Adaptive Control and Signal Processing 22(10), 909–931 (2008)
121. Wu, L., Shi, P., Gao, H., Wang, J.: h_∞ model reduction for linear parameter-varying systems with distributed delay. International Journal of Control 82(3), 408–422 (2009)

122. Wu, L., Zheng, W.: Weighted h_∞ model reduction for linear switched systems with time-varying delay. Automatica 45(1), 186–193 (2009)
123. Wu, M., He, Y., She, J., Liu, G.: Delay-dependent criteria for robust stability of time-varying delay systems. Automatica 40(8), 1435–1439 (2004)
124. Xiang, Z., Qiao, C., Mahmoud, M.S.: Robust h_∞ filtering for switched stochastic systems under asynchronous switching. Journal of the Franklin Institute 349(3), 1213–1230 (2012)
125. Xiang, Z., Wang, R.: Robust control for uncertain switched non-linear systems with time delay under asynchronous switching. IET Control Theory and Applications 3(8), 1041–1050 (2009)
126. Xie, D., Chen, X., Lv, L., Xu, N.: Asymptotical stabilisability of networked control systems: time-delay switched system approach. IET Control Theory and Applications 2(9), 743–751 (2008)
127. Xie, D., Wang, L., Hao, F., Xie, G.: Robust stability analysis and control synthesis for discrete-time uncertain switched systems. In: Proceedings of the 42nd IEEE Conference on Decision and Control, vol. 5, pp. 4812–4817 (2003)
128. Xie, G., Wang, L.: Quadratic stability and stabilization of discrete-time switched systems with state delay. In: 43rd IEEE Conference on Decision and Control, vol. 3, pp. 3235–3240 (2004)
129. Xie, W., Wen, C., Li, Z.: Input-to-state stabilization of switched nonlinear systems. IEEE Transactions on Automatic Control 46(7), 1111–1116 (2001)
130. Xu, S., Lam, J.: h_∞ model reduction for discrete-time singular systems. Systems and Control Letters 48(2), 121–133 (2003)
131. Xu, S., Lam, J.: Exponential h_∞ filter design for uncertain takagi–sugeno fuzzy systems with time delay. Engineering Applications of Artificial Intelligence 17(6), 645–659 (2004)
132. Xu, S., Lam, J.: A survey of linear matrix inequality techniques in stability analysis of delay systems. International Journal of Systems Science 39(12), 1095–1113 (2008)
133. Xu, S., Lam, J., Huang, S., Yang, C.: h_∞ model reduction for linear time-delay systems: Continuous-time case. International Journal of Control 74(11), 1062–1074 (2001)
134. Yan, W., Lam, J.: An approximate approach to h_2 optimal model reduction. IEEE Transactions on Automatic Control 44(7), 1341–1358 (1999)
135. Ye, H., Michel, A., Hou, L.: Stability theory for hybrid dynamical systems. IEEE Transactions on Automatic Control 43(4), 461–474 (1998)
136. Yu, M., Wang, L., Chu, T., Xie, G.: Stabilization of networked control systems with data packet dropout and network delays via switching system approach. In: 43rd IEEE Conference on Decision and Control, vol. 4, pp. 3539–3544 (2004)
137. Zhai, G., Hu, B., Yasuda, K., Michel, A.: Disturbance attenuation properties of time-controlled switched systems. Journal of the Franklin Institute 338(7), 765–779 (2001)
138. Zhai, G., Hu, B., Yasuda, K., Michel, A.: Stability analysis of switched systems with stable and unstable subsystems: an average dwell time approach. International Journal of Systems Science 32(8), 1055–1061 (2001)
139. Zhai, G., Lin, H.: Controller failure time analysis for symmetric control systems. International Journal of Control 77(6), 598–605 (2004)
140. Zhai, G., Sun, Y., Chen, X., Michel, A.: Stability and l_2 gain analysis for switched symmetric systems with time delay. In: Proceedings of the 2003 American Control Conference, vol. 3, pp. 2682–2687 (2003)
141. Zhang, L., Boukas, E., Shi, P.: Exponential h_∞ filtering for uncertain discrete-time switched linear systems with average dwell time: A μ-dependent approach. International Journal of Robust and Nonlinear Control 18(11), 1188–1207 (2008)

142. Zhang, L., Gao, H.: Asynchronously switched control of switched linear systems with average dwell time. Automatica 46(5), 953–958 (2010)
143. Zhang, L., Huang, B., Lam, J.: h_∞ model reduction of markovian jump linear systems. Systems and Control Letters 50(2), 103–118 (2003)
144. Zhang, L., Shi, P., Boukas, E., Wang, C.: Robust $l_2 - l_\infty$ filtering for switched linear discrete time-delay systems with polytopic uncertainties. IET Control Theory and Applications 1(3), 722–730 (2007)
145. Zhang, L., Shi, P., Boukas, E., Wang, C.: h_∞ model reduction for uncertain switched linear discrete-time systems. Automatica 44(11), 2944–2949 (2008)
146. Zhang, L., Shi, P., Wang, C., Gao, H.: Robust h_∞ filtering for switched linear discrete-time systems with polytopic uncertainties. International Journal of Adaptive Control and Signal Processing 20(6), 291–304 (2006)
147. Zhang, P., Ding, S.: An integrated trade-off design of observer based fault detection systems. Automatica 44(7), 1886–1894 (2008)
148. Zhao, J., Hill, D.: On stability, l_2-gain and h_∞ control for switched systems. Automatica 44(5), 1220–1232 (2008)
149. Zhao, X., Zeng, Q.: Delay-dependent h_∞ performance analysis and filtering for markovian jump systems with interval time-varying delays. International Journal of Adaptive Control and Signal Processing 24(8), 633–642 (2010)
150. Zheng, Y., Feng, G., Qiu, J.: Exponential h_∞ filtering for discrete-time switched state-delay systems under asynchronous switching. Asian Journal of Control, 1–10 (2012)
151. Zhong, M., Ding, S., Lam, J., Wang, H.: An lmi approach to design robust fault detection filter for uncertain lti systems. Automatica 39(3), 543–550 (2003)
152. Zhong, M., Ye, H., Shi, P., Wang, G.: Fault detection for markovian jump systems. IET Control Theory and Applications 152(4), 397–402 (2005)

MIX
Papier aus verantwortungsvollen Quellen
Paper from responsible sources
FSC® C105338

If you have any concerns about our products, you can contact us on
ProductSafety@springernature.com

In case Publisher is established outside the EU, the EU authorized representative is:
Springer Nature Customer Service Center GmbH
Europaplatz 3, 69115 Heidelberg, Germany

Printed by Libri Plureos GmbH
in Hamburg, Germany